1n memoriam
Lee Schwarz
1959~1974

INFINITE VISTAS

INFINITE VISTAS

New Tools for Astronomy

Edited by

JAMES CORNELL

and

JOHN CARR

CHARLES SCRIBNER'S SONS

NEW YORK

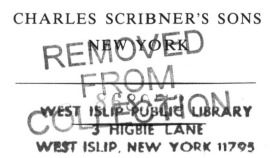

Copyright © 1985 by Smithsonian Institution Astrophysical Observatory

Library of Congress Cataloging in Publication Data

Main entry under title:
Infinite vistas.
Bibliography: p.
Includes index.
Contents: First light, the space telescope / Alan
P. Lightman—Giant telescopes and tall mountains /
David W. Latham—Speckle imaging / Robert
Stachnik—[etc.]
1. Telescope—Addresses, essays, lectures.
2. Astronomical instruments—Addresses, essays,
lectures. 3. Astronomy—Research—Addresses,
essays, lectures. I. Cornell, James. II. Carr,
John.
QB88.I54 1985   522'.2   85-14278
ISBN 0-684-18287-4

Published simultaneously in
Canada by Collier Macmillan Canada, Inc.—
copyright under the Berne Convention.

1 3 5 7 9 11 13 15 17 19 F/C 20 18 16 14 12 10 8 6 4 2

Printed in the United States of America.

Contents

Acknowledgments

In 1982, the National Academy of Sciences issued, for the benefit of scientific planners and policymakers, the *Report of the Astronomy Survey Committee* listing the research priorities for the 1980s and 1990s, as well as the instruments and facilities deemed necessary for accomplishing these goals. The committee, representing the diverse American astronomical community and its many subfields, was chaired by George B. Field, then director of the Harvard-Smithsonian Center for Astrophysics (CFA).

Inspired by the report's catalog of proposed instruments for the future, the CFA's Public Affairs Office used the theme of "New Tools for Astronomy" for a series of six related lectures presented in the spring of 1982 as part of the CFA's regular monthly "Observatory Nights for the Public." The lectures on the technology and techniques of astronomy, somewhat of a departure from the usual public talks on the results of research, proved surprisingly popular.

In the spring of 1984, one of the editors of this volume (Cornell) arranged for the presentation of the same lectures, with the addition of an introductory talk by George Field, in Washington,

D.C., at the Smithsonian Institution's National Air and Space Museum.

Later that same year, the other editor (Carr) coordinated a third presentation of lectures, this time as part of the Lowell Lectures on Astronomy at the Boston Museum of Science. ("New Tools for Astronomy" actually marked the tenth annual series of free popular lectures produced jointly by the Center for Astrophysics and the Museum's Charles Hayden Planetarium. Two previous series also resulted in book publication—*Revealing the Universe* [Cambridge, Mass.: MIT Press, 1982]; and *Astronomy from Space* [Cambridge, Mass.: MIT Press, 1983].)

The largest part of this volume, then, is drawn from the lectures originally given by seven speakers: David Latham, Robert Stachnik, George Withbroe, Mark Reid, Wesley Traub, Martin Zombeck, and George Field. To fill in the gaps along the electromagnetic spectrum or to describe other proposed instrumentation, three authors prepared chapters especially for this work: Steven Willner, Trevor Weekes, and Paul Ho. And, Alan Lightman adapted and updated the contribution on the Hubble Space Telescope that had originally appeared in his book *Time Travel and Papa Joe's Pipe* (New York: Scribners, 1984).

The editors are grateful for the help of many people in the effort first to bring this information to the lecture hall and then to convert it into book form. The original lectures were conceived by Arlene Walsh of the Center for Astrophysics. At the National Air and Space Museum, Director Walter Boyne enthusiastically endorsed the idea of a joint lecture program; Von Del Chamberlain made the initial arrangements, which were later completed and coordinated by David DeVorkin; and Rita Cipalla prepared publicity and public information materials. The travel of the CFA astronomers based in Cambridge, Massachusetts, to the Washington lecture hall was provided by a grant from the James Smithson Society of the Smithsonian Institution.

At the Boston Museum of Science, the encouragement of Director Roger Nichols was no less invaluable than the logistical support provided by Tina Pala and Nancy DiCiaccio of the Planetarium staff. Valuable technical assistance came from Ray Crane, Val Wilcox, and Joyce Towne. (Astronomer Daniel Schwartz of the CFA also gallantly stepped in at the last moment to present the Boston lecture on X-ray astronomy when colleague Martin

Zombeck was called to Europe.) The series itself would not have been possible without the generosity of John Lowell and the Lowell Institute. The continued support of these popular lectures for the people of Boston is deeply appreciated.

The pains of manuscript preparation were eased considerably by the staff of the CFA's Publications Department. In particular, Joseph Singarella and John Hamwey prepared many of the illustrations for this book, Charles Hanson converted innumerable slides, drawings, and sketches into reproducible photographs, and Anne Omundsen provided expert copy editing. Finally, even in an age when most typescript for a book of this sort comes from the electronic pulses of word processors, it is reassuring to know that excellent typing—not to mention thoughtful formatting, spelling corrections, and personal attention to detail—is still available from Gerda Schrauwen and Mary Juliano.

—JAMES CORNELL and JOHN CARR
Boston

Introduction

He went down the spectrum, noting the evidence of visible heat die out on the scale of the instrument until he came to the apparent end even of the invisible beyond which the most prolonged researches of investigators up to that time had shown nothing . . . By some happy thought he pushed the indications of this delicate instrument into the region still beyond. In the still air of this lofty region the sunbeams passed unimpeded to the mists of the lower earth, and the curve of heat, which had fallen to nothing, began to rise again. There was something there. For he found, suddenly and unexpectedly, a new spectrum of great extent, wholly unknown to science and whose presence was revealed by the new instrument, the bolometer.

—1900 *Annual Report*
Smithsonian Institution

In 1881, the American astrophysicist Samuel Pierpont Langley took a new and unusual instrument of his own invention to the upper reaches of Mt. Whitney in California to measure an elusive quality of sunlight known as the "solar constant." Surprisingly,

Langley's crude but imaginative heat detector—the bolometer—
not only advanced his original goal but serendipitously revealed a
previously unknown region of the solar spectrum, the longer infrared wavelengths.

Although extraordinary, Langley's experience was by no means
unique in science. Indeed, ever since 1610 when Galileo Galilei
turned his rough optical telescope on the heavens and found that
the vague cloud known as the Milky Way was, in fact, "a congeries
of innumerable stars grouped together in clusters," the advance
of astronomical discovery has followed closely the development of
new instrumentation. Think, too, of Karl Janksy's accidental discovery of radio-wave emission from the heavens. Or, how George
Ellery Hale's giant optical reflectors resolved the host of vague,
fuzzy "nebulosities" into external galaxies, "island universes" like
our own.

This pace of technology-precipitating discovery has been most
dramatic in the past quarter-century as the advance of space flight
opened new ultraviolet, gamma-ray, and X-ray windows on the
universe. On the ground, too, the development of highly efficient
telescopes and detectors, enhanced by computer-processing techniques, has expanded the view in the optical, infrared, and radio
wavebands. Indeed, today's astronomers can routinely observe the
heavens across virtually the entire electromagnetic spectrum. And
the ability to combine and correlate observations of the same object
in several different wavelengths has led to a better understanding
of the physical processes governing all matter, from the stars to
the molecules of life.

Although the direct relationship between new instrumentation
and new discoveries is clearly recognized, it is no longer practical—
or even possible—for the visionary scientist simply to patch together magnifying lenses in a wooden tube, walk into the evening
dark, and discover unknown worlds. Not only have all the "easy"
tasks of astronomy been accomplished, but society itself has become more complicated. In the late twentieth century, astronomy—all science, really—is no longer so much an individual en-
terprise as a collective activity, supported by the general public,
responding to national goals, and answering broad questions.

More practically stated, the advance of modern astronomy—
through the development of new instrumentation—now requires
copious funding, large teams of specialists, and, most important,

many years of careful planning and design. Indeed, the time scale for most major instruments is a decade or more, especially if the instrument is to be a national or international facility.

This volume describes the advanced astronomical instruments proposed for the future, instruments intended to extend our vision to some of the most distant objects in the cosmos and to enhance our ability to understand the physical processes at work in these sometimes exotic objects. Unlike the other sciences, astronomy is almost totally observational; yet, aside from a few lunar samples and the occasional meteorites that reach Earth, the astronomer's primary observational data are the faint radiation from stars and galaxies that has traveled millions of miles—and millions of years— through an intervening and obscuring medium before it reaches Earth. Indeed, only a very small portion of the broad and marvelously diverse spectrum of celestial radiation even penetrates the thick and murky blanket of atmosphere overhead. (Figure 1.)

Not surprisingly, then, every astronomer since the time of Galileo has tried to improve the view of the heavens. One approach has been to open new windows in the electromagnetic spectrum. This has been partially achieved by sending detectors above the atmosphere in rockets, balloons, and spacecraft to observe wavelengths blocked from ground-based telescopes. A second approach has been to improve the resolution of astronomical images by creating larger apertures. (Figure 2.)

Of course, practical considerations of weight and cost limit the size of any single mirror or antenna; and, despite the obvious advantages of carrying telescopes into space, the costs and complexity of launching and then operating sensitive instruments in orbit limit the number of experiments possible. As a result, astronomers have pursued alternative and innovative means of achieving better resolution. For example, one way to increase aperture size is by employing the principle of interferometry, in which the signals gathered by two or more telescopes are combined to produce a resolution equivalent to that of a single instrument with a diameter equal to the maximum distance between the two telescopes.

Readers will quickly discover that the theme of improved resolution, and the attempts to achieve it through observations from space platforms or with interferometric techniques, is a prevalent one in this book. Another general, and perhaps obvious, theme

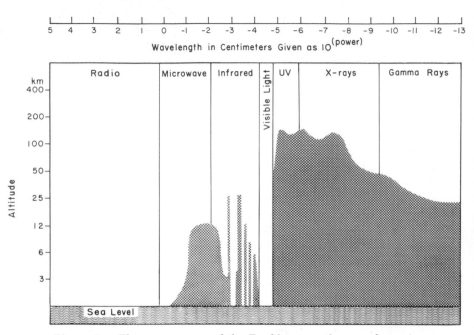

Figure 1: The transparency of the Earth's atmosphere to electronic radiation from space. At each wavelength, the shaded region shows the altitude range over which starlight of that wavelength is effectively "blocked out." Here more than half the radiation is lost to absorption by air. Only above the shaded region can most of the radiation be detected. For example, in the far-infrared (10^{-2} centimeter wavelength), instruments must be sent to an altitude of at least 15 kilometers (60,000 feet) in order to be able to make useful measurements. (Smithsonian Astrophysical Observatory illustration by Joseph Singarella)

is how an incredible rush of new ideas, new concepts, new insights, and, sometimes, new questions about the universe follows the opening of each new wavelength window.

What may be surprising, in this age of satellites, is how much astronomy will still be done from the ground. Although the Hubble Space Telescope, now scheduled to be launched shortly after this book is published, has been touted as the ultimate instrument for modern astronomy and astrophysics, it is only the first in a coordinated arsenal of both orbiting and ground-based observatories, including others for radio, X-ray, infrared, and gamma-ray astronomy. Moreover, the presence of a precision optical telescope

APERTURE SIZE AND ANGULAR RESOLUTION

INSTRUMENT	DIAMETER (meters)	COLLECTING AREA (square meters)	ANGULAR RESOLUTION (arcsec)	SEPARATION (meters)
Human Eye	0.0066	0.00003	120	240,000
Ground-Based Optical Telescope	0.1–6.0	0.008–28.0	1	2,000
VLA (radio interferometer)	36,000	13,000	0.05	100
Space Telescope	2.4	4.5	0.05	100
Speckle Imaging (ground-based telescope)	5	20	0.02	40
COSMIC (space optical interferometer)	36	45	0.003	6
VLBA (radio interferometer)	8,000,000	5,000	0.0002	0.4
Space VLBI	50,000,000	5,000+*	0.00005	0.1

*If used in conjunction with the ground-based VLBA.

Figure 2: *Astronomical measurements are usually expressed as degrees, minutes, and seconds of an angle of arc on the celestial sphere; for example, the diameter of the Moon's disk measures one-half degree of arc—or 30 arcminutes—as seen from Earth. A telescope's angular resolution, that is, its ability to measure the diameter of a distant object or to distinguish between two adjacent objects, generally increases with the size of the aperture. The ability of an instrument to detect faint objects generally increases with its collecting area. In the dark of night, the human eye has about seven times poorer angular resolution than might be expected from diffraction theory. The angular resolution of ground-based telescopes at visible wavelengths is typically about 1 arcsecond and is limited by atmospheric turbulence. The resolution of radio interferometers is determined by the distance between the two most widely separated receivers in an array. The table above shows the relative diameters and collecting areas of existing and proposed instruments, with the resultant resolving power of each. The table also shows how far two features on the Moon's surface would need to be separated for them to be seen as two distinct objects by an observer on Earth using the different instruments.* (Figure compiled by Wesley Traub and Mark Reid)

in space and its almost inevitable discovery of new phenomena can only increase the demand and need for other instruments on Earth.

There is another theme, or attitude, in this particular book that requires some explanation—or warning—for the reader. To achieve significant improvements over the tools of the past, the next generation of instruments will require increasingly sophisticated and complex technical advances. By definition, then, many of the instruments and concepts described here are based on advanced electronic systems, subtle mathematical principles, and complicated operating procedures. This book is not so much concerned with the stunning or beautiful results of astronomical observations as with the incredible means by which these results are achieved. Although the use of mathematical symbols and equations has been limited, it has not always been possible to avoid technical descriptions of facilities, systems, or procedures. In part, this is intentional: to impress upon the reader just how difficult and demanding are the tasks of modern astrophysics. But it has also been assumed that a goodly part of the public is now also technically sophisticated—members of a society in which computers, electronics, and technical devices are commonplace and, if not totally understood, certainly not feared. Fortunately, the authors of the different chapters are as varied in their style and approach as in their subject matter; thus, readers should find much that matches their own levels of interest or technical expertise.

Finally, readers should be aware that only a few of the new tools for the future described here are so far certain of construction. Most ultimately depend on federal funding; and, while American astronomers have proposed a comprehensive, coordinated program for the next two decades, it is not certain that Congress will share their vision. Obviously, then, the instrument-development programs described here require not only imagination, innovation, and ingenuity, but considerable institutional courage as well. Risks are inherent in all pioneering attempts, but the scale of many of these instruments suggests that time, money, careers, and individual reputations must be committed many years in advance to projects whose outcome cannot be guaranteed or, in some cases, even imagined. Perhaps it is no coincidence that all the authors, and many of the projects they describe, are located at the Harvard-Smithsonian Center for Astrophysics. Both the Harvard College Observatory and the Smithsonian Astrophysical Observatory, the two independent research institutions that form the Center, have long histories of such scientific risk-taking.

It was at Harvard, for example, that the earliest photographic plates of astronomical objects were made, thus sparking the first major revolution in the way data would be taken and stored for astronomy. More recently, in the early 1960s, the Harvard Solar

Satellite Project developed some of the first space telescopes—at a time when most traditional astronomers were still wary of placing their hopes and dreams in the nose cones of undependable and unpredictable rockets.

At the Smithsonian, the tradition of innovation in engineering and instrument making can be traced to the 1880s and S. P. Langley's bolometer. In the late 1950s, Smithsonian astronomer Fred Whipple created an unusual tracking camera that stood ready to photograph the first artificial satellites. And, in the 1970s, Smithsonian astronomers joined with colleagues at the University of Arizona to create the Multiple Mirror Telescope. Considered radical, revolutionary, and, to some, even foolhardy when first conceived, the idea of multiple-mirror arrays has since been recommended as the preferred design for the proposed National New Technology Telescope, a 15-meter optical giant that will become the world's largest telescope when completed sometime in the next decade.

Indeed, one of the intangible benefits of this book may be its function as prophecy. Perhaps the astronomers of the next century will look back at its predictions as guideposts for the advance of modern astronomy.

—JAMES CORNELL

INFINITE VISTAS

First Light

The Space Telescope

ALAN P. LIGHTMAN

Sometime in 1986 the cargo bay of the Space Shuttle will open, 300 miles above the Earth, and a mechanical arm will release into orbit a 42-foot-long cylinder containing a telescope. The Hubble Space Telescope (Figure 1.1), as it is called in honor of the American astronomer Edwin P. Hubble, has been twenty years in the planning and will cost the National Aeronautics and Space Administration and the European Space Agency over a billion dollars. It is the first—and only certain—new tool for astronomy in the next decade. It will also certainly dominate astronomical research for the rest of the century.

Why send a telescope into space? Because the Earth's atmosphere, while quite agreeable to most of us, is a headache to observational astronomers. First and foremost, the images of astronomical objects are blurred when light travels through the turbulent and clumpy air around the Earth: that's why stars twinkle. Beneath the murky atmosphere, ground-based optical telescopes cannot normally distinguish details separated in angle by less than about three ten-thousandths of a degree (the half sky has a total angle of 180 degrees). The Hubble Space Telescope will be able to see ten times more clearly than this—clearly enough, for example, to

Figure 1.1: *Artist's concept of the Hubble Space Telescope in orbit. The aperture door (right end) is open. Rectangular solar arrays extending from both sides of the spacecraft help supply power. The high-gain antenna extending above the telescope perpendicular to its long axis is one of two (the other, unseen here, extends below) that receive and transmit data through* NASA's Tracking and Data Relay Satellite System. (NASA illustration)

read the license plate of a car in Boston from as far away as Washington, D.C. (Figure 1.2a and b.)

Ground-based telescopes also receive stray light from cities and from atmospheric auroras. Background light becomes confused with light from the object under study. Such unwanted contamination almost disappears high above the Earth's atmosphere. At visible wavelengths of light, outer space is about three times darker than earthly nights in most locations. With its decreased background light and increased angular resolving power, the Hubble Space Telescope will be able to see stars fifty times fainter than those observed by ground-based telescopes now in use.

Finally, atmospheric absorption prevents much of the radiation produced by astronomical bodies from ever reaching Earth. This includes ultraviolet and infrared radiation, with wavelengths respectively shorter and longer than those of visible light. With appropriate instruments, the Hubble Space Telescope will operate over several hundred times more of the electromagnetic spectrum than can comparable telescopes here on Earth.

Dreams of doing science in space are not new. More than a century ago, Jules Verne imagined exploration of Africa by balloon in his *Five Weeks in a Balloon*, published in 1863, and described man-made satellites in *From the Earth to the Moon* (1865). The German rocketry pioneer Hermann Julius Oberth was one of the first to point out that space is where telescopes ought to be. That was in 1923. The first astronomical observations in space were made in the late 1940s, using captured German V-2 rockets that were capable of poking above the atmosphere for a few minutes. In the late 1950s, a U.S. telescope named Stratoscope I was lifted to the top of the atmosphere by balloon. The first orbiting astronomical satellites, with the precious advantages of relative stability and longer lifetimes, were operated by NASA in the 1960s.

In 1978, the United States placed two major astronomical satellites into orbit—the International Ultraviolet Explorer and the Einstein Observatory. The latter, at a cost of about $200 million (in 1985 dollars), housed the first telescope able to focus X rays, radiation of even shorter wavelength than the ultraviolet. Einstein ran out of gas in April 1981. More precisely, the small jets on the satellite, crucial for changing its orientation on command, ran out of gas. The Hubble Space Telescope, which will be serviced regularly by the Shuttle, should enjoy a lifetime of at least 15 years.

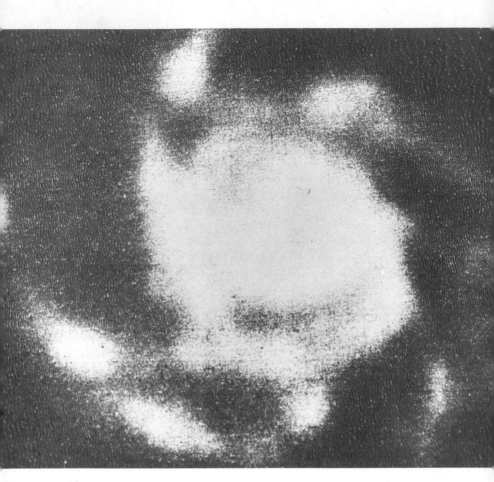

(a)

Figure 1.2: *(a) A visible light image of a distant spiral galaxy. (b) The same spiral galaxy seen by a telescope with a resolution 10 times better.* (NASA illustration)

It is the first major scientific space project since the Einstein Observatory.

As an illustration of the power of the telescope, consider its potential impact on several fundamental questions that keep astronomers up at night. One such question is whether there is any life in the universe beyond our own planet. At one time the chances

(b)

of locating life forms on sister planets in our solar system seemed good. Unfortunately, the Viking landers that touched down on Mars found no signs of life. And Mars had been considered the best bet. What about planets circling stars other than the Sun? There are one hundred billion stars just in our own galaxy, the Milky Way. Perhaps with the billions of possible temperatures, gravities, and chemical compositions on other worlds, life might flourish somewhere. But astronomers thus far have found only a single candidate planet outside our solar system, about 21 light-years away. And the question of whether this object is really a planet or small dim star (brown dwarf) is still controversial. Present ground-based telescopes have great difficulty in distinguishing the relatively faint reflected light of a hypothetical planet from the much brighter light of its parent star, or discerning the tiny periodic wobble of the parent star in response to the gravity of an orbiting planet. Of course, finding other planetary systems in the universe isn't equivalent to documenting the footprints of extraterrestrials, but it may be an important first step.

For example, the star nearest the Sun is Alpha Centauri, about 4.5 light-years distant. Alpha Centauri is not unlike our own Sun. If it had a planetary system similar to our own, its position would not stay fixed but would shift back and forth by about a millionth of a degree. (Figure 1.3.) Ground-based telescopes are unable to detect such a small movement. The Hubble Space Telescope, however, should be able to sense the wobble of planet-bearing stars as far away as 10 or 20 light-years, and will be programmed to scrutinize a dozen or so candidates. Theoretical astronomers, who work only with pencil and paper, would be startled if space were not littered with planetary systems. Nevertheless, the question of our uniqueness seems sufficiently unsettling that it would be nice to know for sure.

How old is the universe? Since the 1920s we have known that the universe is in a state of expansion, with the galaxies rushing away from each other. If this moving picture is mentally played backward in time, then the galaxies crowd ever closer together until a definite instant in the past when all matter in the universe crushes together into a point of unimaginable density. Almost all astronomers and physicists accept the concept that the universe began with the Big Bang, when that matter started the expansion we still see. How long ago was it?

Since all galaxies originated from the same point in the Big Bang, the time elapsed since then is the time required for any galaxy some distance from the Milky Way to move to wherever it is now. Determining other galaxies' velocities and distances from our galaxy is therefore crucial. For example, if the galactic velocities were constant in time, a galaxy flying away at 0.1 billion light-years per year would take 10 billion years to put one billion light-years between itself and us. Astronomers have no trouble measuring the velocities of nearby galaxies to high accuracy, but their distances from us are currently known to only about 30 percent. In terms of the birthday of the universe, this translates to an uncertainty of several billion years. (Figure 1.4.)

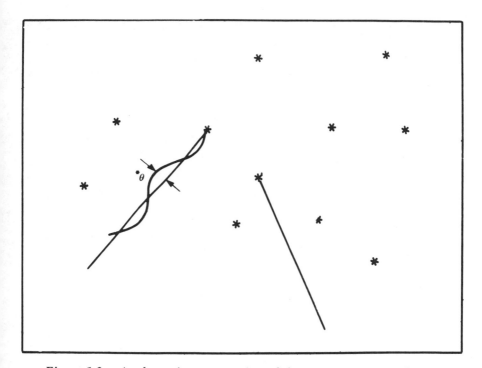

Figure 1.3: *A schematic representation of the apparent motion of two stars. One appears to move in a straight line among the background stars; the other appears to move in a sinusoidal pattern. Because the center of mass of a system always moves linearly, a star that wobbles must have an unseen companion. (*Smithsonian Institution illustration)

Measuring cosmic distances is like climbing an endless ladder into space, where the size of each step is estimated by the height of the previous rung. To begin, you must somehow determine the distance to a nearby standard object, like a type G8 V star. All G8 V stars have the same intrinsic properties, including luminosity (absolute brightness). By comparing the *apparent* brightness of the type G8 V star with its distance from the observer, you can infer its luminosity. (The apparent brightness of a light source of given luminosity decreases as it moves farther away from the observer.) With G8 V stars in hand for calibration, you climb one rung up the ladder and seek out a G8 V star harbored within a brighter, more distant object—say a large globular cluster of stars. From the apparent brightness of the new G8 V star, whose luminosity

was previously determined, you can judge the distance to the globular cluster; you can then calculate the luminosity of the cluster
from *its* apparent brightness. In the process you have happily gained
a new type of object for calibration, the globular cluster, and can
now proceed onward up the ladder, perhaps looking for globular
clusters embedded within galaxies. Needless to say, the higher you
go up the ladder, the shakier it gets.

The first rung had better be very trustworthy. As Earth revolves
around the Sun and changes its perspective, the apparent positions
of nearby stars shift on the night sky, providing an estimate of
their distances (closer stars shift more). The Hubble Space Telescope, with its highly precise angular resolution, should be five
times more accurate than ground-based telescopes in measuring
the distances to nearby stars in this manner. For the next couple
of rungs, taking us out to nearby galaxies, the Hubble Space Telescope should obtain improved measurements of certain stars used
for calibration. Several rungs up the cosmic distance ladder and
on solid footing, we may be able to determine the rate of expansion
of the universe, and ultimately its age, to within several percent.

Do black holes exist? On paper, of course they do. However,
aside from one strong candidate known as Cygnus X-1, about 7,000
light-years from Earth, astronomers have floundered in their search
for these exotic objects. According to theory, black holes, which
are collapsed stars, should have a very small size. (A black hole
of the same mass as our Sun would be 1.5 miles in diameter; larger
black holes would have diameters in proportion to their mass.)
There is a clever way of measuring such small sizes: all objects,
on Earth or in space, are continually subject to disturbances, which
produce variations in the emission of light and sound. In astronomy, the size of an object too distant to make out directly may be
inferred from the rapidity of its light fluctuations, in much the
same way that the size of a canyon can be estimated by the interval
between echoes. An instrument aboard the Hubble Space Telescope will be able to distinguish light fluctuations spaced as closely
as ten millionths of a second apart. Such a time interval would

Figure 1.4: *Galaxies as seen in a wide-field photograph by the Mayall 4-
meter telescope.* (Kitt Peak National Observatory photograph)

indicate an object whose size is ten millionths of a light-second, or about two miles. Most black holes would be no smaller.

Another strategy for unmasking black holes is less direct. A massive black hole inhabiting the center of a galaxy, while busily chewing up stars, should force the surviving stars to huddle around it. The hapless stars queuing up for destruction should appear as a particular increase in light toward the center of the galaxy. So far this diagnostic effect has escaped detection by earthly telescopes. Again it is a matter of angular resolution. For galaxies within two million light-years of Earth, including Andromeda and several others, the Hubble Space Telescope should be able to peer within one light-year of the galactic center, perhaps close enough to see the handiwork of a black hole 100 million times the mass of the Sun.

To carry out its awesome business, the Hubble Space Telescope will require an impressive armament of on-board instruments, technical innovations, and organizational support on the ground. First, there is the telescope itself, which will have a 2.4-meter (94-inch) primary mirror to focus incoming light, built by the Perkin-Elmer Corporation. (Figure 1.5.) For analyzing the light, six scientific instruments are now scheduled to be on board the satellite: the wide-field/planetary camera, the faint-object camera, the faint-object spectrograph, the high-resolution spectrograph, the high-speed photometer, and the fine-guidance system. Each has been built by a team of scientists and will have required more than 10 years to complete by the time the telescope is launched.

A terrestrial photographer will be surprised not to find film in the Hubble Space Telescope. The two cameras record light intensities with electronic detectors known in the trade as CCDs (charge-coupled devices). Incoming light is broken up into its constituent colors by the two spectrographs, permitting such details as the temperatures and chemical compositions of objects in space

Figure 1.5: Technicians inspect the primary mirror for the Hubble Space Telescope. The mirror's aluminum-magnesium-fluoride coating covers a titanium-silicate glass core constructed with a strong and lightweight honeycomb configuration. Measuring 2.4 meters (94.5 inches) in diameter and weighing 829 kilograms (1,827 pounds), the mirror will stay at a nearly constant temperature in orbit to avoid distorting surface changes. (NASA photograph)

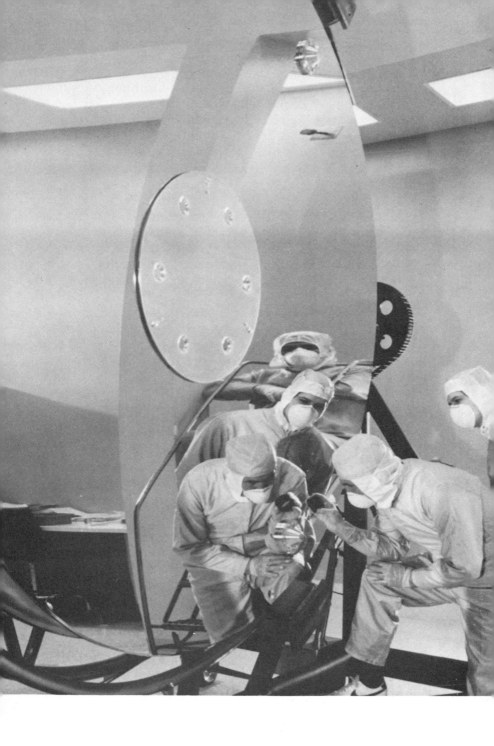

to be deduced. The high-speed photometer measures the variability in the intensity of light and produces data that can be used, among other things, to infer the sizes of objects. Finally, the fine guidance system, with its gyroscopes and star indentification systems, will hold the telescope's direction steady to three millionths of a degree over a period of 10 hours, or a little more than six orbits.

Capitalizing on the technical opportunities in space has required breakthroughs in materials and design. A good example is the telescope's support structure, which holds the primary and secondary mirrors apart at a distance of 17 feet. Temperature changes in space could cause distortions in the structure, and expansion or contraction of as little as one ten-thousandth of an inch would

Figure 1.6: Artist's depiction of the optical telescope assembly (OTA) and scientific-instruments section of NASA's Hubble Space Telescope. The instruments, designed as modular units, are shown to the right, behind the 2.4-meter primary mirror. (NASA illustration)

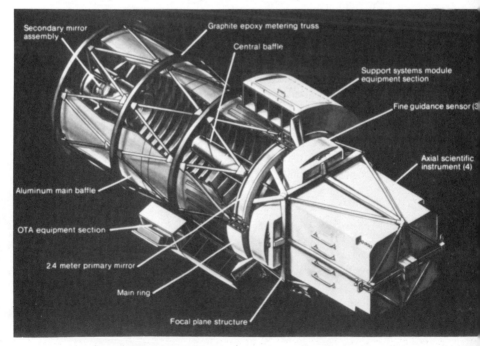

produce tears in the data rooms on Earth. To solve the problem, Boeing Aerospace has developed for the structure a mixture of graphite, which expands with cold and contracts with heat, and epoxy, which expands with heat and contracts with cold. (Figure 1.6.)

Astoundingly, all this expensive hardware and technology is going to be shot off into space, largely unattended and inaccessible. An on-board computer will control the program of observations. Communication with the telescope, including data retrieval, will occur via a pair of satellites, called the Tracking and Data Relay Satellite System (TDRSS), which are already hovering over fixed positions on the Earth's surface. On board, all data are digitized (converted into bits of information represented by ones and zeros), and can be stored on magnetic tape before radio transmission to the TDRSS satellites and then earthward to the Goddard Space Flight Center in Maryland—and human beings. (Figure 1.7.)

Observational astronomy, these days, like many other fields of experimental science, has taken on a ghostly quality. It used to be that astronomers on "observing runs" would pack up several days of sandwiches and good books for the cloudy nights, travel to the top of a mountain somewhere, and sit at the eyepiece of a telescope, taking notes and photographs and simply enjoying the spectacle firsthand. A colleague of mine, working with data from the Einstein X-ray satellite, recently completed a "hands-on" investigation of quasars. When I asked him what it was like, he allowed that in fact he had passed the time in front of a video screen just down the hall, pushing keys and pondering over various digitized images of the quasar from data stored on magnetic tape. The information had previously been manipulated by two other computers, before which it had been telemetered from space to Earth in digitized form. It was irrelevant that the Einstein satellite, which alone "saw" the quasar, had been defunct for several years. Digitized data keep well.

And there will be enormous quantities of data from the Hubble Space Telescope. To digest such an onslaught of cosmic information, a new facility, the Space Telescope Science Institute, has recently been established on the campus of Johns Hopkins University in Baltimore. The institute, as well as the observing program of the telescope, will be administered for NASA by the Association of Universities for Research in Astronomy, a consortium of sev-

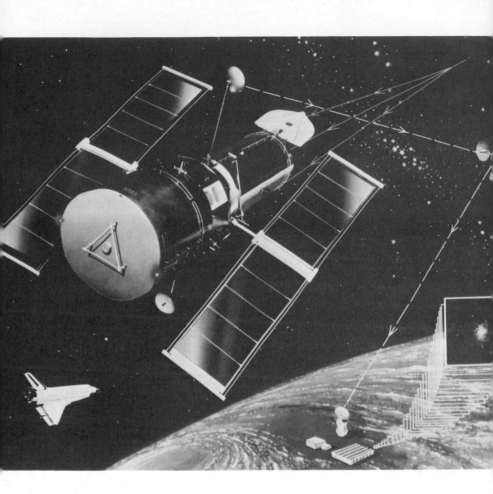

Figure 1.7: *Artist's schematic showing communications with the Hubble Space Telescope. Light enters the aperture, strikes the primary mirror, reflects back to the secondary mirror, then reflects through a hole in the primary mirror to the focal plane. Scientific instruments collect their information at the focal plane. Two high-gain antennas extending from the telescope perpendicular to its long axis transmit this information to a NASA Tracking and Data Relay Satellite (TDRS)—shown, not to scale, at the picture's upper right. The telescope orbits in a Shuttle-serviceable 300-mile orbit, while the TDRS uses a geosynchronous orbit, around 22,700 miles. The TDRS transmits the information to NASA's ground station at White Sands, New Mexico, which in turn feeds the data via commercial satellite to the Goddard Space Flight Center at Greenbelt, Maryland, which feeds it by land to the Space Telescope Science Institute in Baltimore. (NASA illustration)*

enteen universities. The staff of the institute may eventually number more than 250 people, including 100 Ph.D.'s, comparable to or larger than the largest national facility now in operation for ground-based optical telescopes, Kitt Peak National Observatory in Tucson, Arizona. It is estimated that the yearly maintenance and operation cost of the telescope and the institute will be about $150 million, or nearly five times the total capital cost of a state-of-the-art optical telescope on the ground and somewhat more than the yearly budget of Kitt Peak. An additional data center, at the European Southern Observatory in Munich, will be administered by the European Space Agency, which is contributing 15 percent of the cost of the telescope.

The Hubble Space Telescope represents an increasing trend in experimental science toward complex instruments, big bucks, and large teams. Where this will lead no one knows, but most scientists view the development as an inevitable consequence of merging new technology with science. Long before the industrial revolution, Francis Bacon shrewdly predicted such a merger. In his *The New Atlantis*, published in 1627, he describes a utopian institution in which "we procure means of seeing objects afar off . . . We have also sound-houses, where we practice and demonstrate all sounds, and their generation. We have certain helps which set to the ear do further the hearing greatly."

The evolution toward big science is nowhere better illustrated than in the progress of telescopes, in which the Hubble Space Telescope stands as a crowning achievement. Telescopes were first applied to astronomy, beginning in 1610, by Galileo, who discovered mountains on the Moon and moons around Jupiter. In 1985 dollars, Galileo's telescope (minus his signature) would cost about $100. While the ability to make bigger and optically better telescope lenses and mirrors steadily increased, new gadgets were developed for light analysis and operation. Spectrographs and photographs were introduced in the 1800s, and photometers came into wide use in the 1930s. Modern computers bounded onto the stage with the invention of the vacuum tube in 1945, were made smaller and faster with the transistor in 1958, and became still smaller and faster with the silicon-chip integrated circuit in 1966.

Recent ground-based optical telescopes, like the 4-meter (157-inch) telescope at Kitt Peak, built in 1970, and the 4-meter Anglo-

Australian Telescope, built in 1974, cost about $30 million each, again in 1985 dollars. The new Multiple Mirror Telescope on Mount Hopkins, Arizona, which has the novel design of six mirrors combined to produce one image with an equivalent mirror aperture of 4.5 meters, cost only $10 million in 1979. At a cost of about $1.4 billion, the Hubble Space Telescope will be definitely out of sight.

Or consider the parallel trend that has occurred in various areas of physics—for example, in the search for smaller and smaller structures. At close range, the naked eye can distinguish structures a few thousandths of a centimeter in size, about the diameter of a human hair. The first microscope, constructed out of cardboard, wood, and vellum by Anton van Leeuwenhoek in the seventeenth century, brought bacteria into focus. It got down to sizes of less than one ten-thousandth of a centimeter, at a cost of about $100 in today's dollars. In recent searches for subnuclear structure, complexities and costs have soared. Dissecting matter, or whatever you call it at a millionth of a billionth of a centimeter, now requires the leviathan "particle accelerators," like Fermilab in Illinois, Brookhaven in New York, and CERN in Switzerland. Fermilab, built in the late 1960s for several hundred million dollars, winds four miles around.

Other fields of basic, experimental science have not yet become consumed by such spectacular machines as the Hubble Space Telescope and the Fermilab accelerator, but they are drifting in that direction. None of this should quite surprise anyone. Scientists have grown accustomed to snooping around in territory far removed from human sense perception. As each science relies more and more on artificial eyes and ears, the snooping will cost more. Perhaps more important, these trends are rapidly separating the objects of science from the scientists studying them. Theoretical science has always been abstract, with its disembodied equations. Now experimental science, in this age of high technology, is finding its own type of abstraction: grasping the spirals of DNA with the reflections of X rays; visualizing the structure of protons from the readouts of machines that collect particles produced by particles produced by protons in the vacuum bowels of giant accelerators; taking the measure of distant, unseen galaxies with instruments that see in the infrared. We speak confidently of a physical reality that we no longer directly see or feel. We put greater and greater faith in our instruments.

Another consequence of the mounting complexity and cost of doing science is that groups are replacing individuals. Although there are exceptions, the lone scientist sounding out nature in a personal lab, with homemade equipment, has become a memory. Says physicist Norman Ramsey of Harvard, "By and large, it's easier to invent a fundamental experiment that takes a big group. Otherwise, it would have been done already." Between the first five years (1958–1962) of the physics journal *Physical Review Letters* and the most recent five years, the average number of authors per experimental paper has increased from 4.6 to 7.0.

Furthermore, a healthy fraction of scientists are now working at large, federally funded national centers similar to the nascent Space Telescope Science Institute. These centers—the National Radio Astronomy Observatory in Charlottesville, Virginia, the NASA Goddard Space Flight Center in Greenbelt, Maryland, the Los Alamos Scientific Laboratory, in New Mexico, and Brookhaven National Laboratory, to name a few—now include 55 percent of all astronomers and 45 percent of all physicists. Los Alamos alone has a staff of about a thousand physicists. To thrive in this kind of environment, the ambitious experimental scientist of today must be much more than a scientist. He or she must also be a manager, an organizer, a grant getter, an entrepreneur.

One wonders whether, in the caverns of big science, some of the bright youngsters might lose their way. Riccardo Giacconi, the director of the Space Telescope Science Institute, believes that we should not place too much emphasis on the issues of depersonalization in modern science. "Although it is true that success requires a chain of command," he says, "it is not clear that most people should be doing independent research." But Giacconi adds: "On the other hand, I am worried whether you select against creativity and individuality by funding large projects."

Samuel Ting, professor of physics and winner of the 1976 Nobel prize, agrees. "The trend toward big science is very unfortunate," he says. "Particularly in experimental particle physics, with its teams of thirty to three hundred people, a young physicist has an extraordinarily difficult time in showing himself. When I started twenty years ago, a particle physics experiment could still be done by four or five people. If I were to start again today, I would not go into this field."

Scientists have traditionally liked to work alone. It's difficult to foresee what will happen when they can't. Will many bright new-comers avoid experiment in favor of theory, which is largely im-mune to rising costs of equipment and still accessible to the in-dependent scientist? Will fundamental science become more like engineering or industry, with a corresponding change in goals and motivation? Will the training of scientists shift from the apprentice-type relationship between thesis advisor and student to the team environment found during the construction of large new instru-ments?

Ironically, the fundamental discoveries have so far been made by relatively small projects and small groups. In his book *Cosmic Discovery*, Martin Harwit, an astronomer at Cornell, concludes that none of the 43 most important astronomical phenomena, in-cluding 14 discovered since 1960, were originally identified at large national centers. In part, this simply reflects the greater time for the use of instruments allotted to individual scientists at private centers of research. The importance of independence and person-alization has not yet been quantified.

I will hazard the guess that many of the major discoveries of the Hubble Space Telescope will be made by small teams not directly associated with the Space Telescope Science Institute, perhaps not even looking for anything specific. Besides the regular investiga-tions of the telescope, there is a wonderful category known as the serendipity mode, consisting of secondary observations made when observing time becomes unexpectedly available. For the big ma-chines, just as for the human mind, discovery may hinge upon unscheduled operations.

Giant Telescopes and Tall Mountains

New Directions in Optical Astronomy

DAVID W. LATHAM

The first time I visited a real astronomical observatory was a frigid December night in 1962. At the door to the 61-inch reflector I was intercepted by a mildly annoyed astronomer, who came stomping out to see who was being so careless with car headlights. To my astonishment, an electrical cord trailed out of a flap in the rear of his heated suit and onto the floor behind him. (Later I learned that even his socks were wired!) When he saw I was just one of the new graduate students, he turned and climbed back up to his precarious perch on the observing platform near the top of the telescope, some 30 feet or so above the floor. There he spent the rest of the night, eye glued to freezing eyepiece, as he made practiced adjustments to the plateholder every minute or so in order to keep the stellar images positioned steadily throughout each long photographic exposure.

The classical picture of an observational astronomer struggling against the elements to spend long hours at the eyepiece of a telescope on the top of some remote mountain peak was valid for almost 100 years. But, in the time since Sputnik, this popular image has been made obsolete by a revolution in electronics. The modern observer spends her nights in a heated room separated from the dome chamber, talking to the computers that run the instruments and detectors and point the telescope. Facing the observer is a bank of television screens for monitoring the various systems. She may pass a whole night without a glimpse of the sky, unless there is a need to check how bad the clouds are. Indeed, modern telescopic observing can resemble the operation of a space satellite, even to the point where the telescope is controlled remotely from a room located thousands of miles away.

In recent years, observational astronomy also has expanded into new regimes, so an astronomer working with a ground-based telescope in visible light is now only one part of astronomy. Many examples of the new tools that are producing sharper and more detailed astronomical images, and that are opening additional regimes of the electromagnetic spectrum, are described in subsequent chapters. In this chapter, I will talk about the changing— but continuing—role of optical telescopes in astronomy, with. emphasis on the recent surge of activity that may produce a new generation of giant telescopes on tall mountains.

In the early years of the twentieth century, preeminence in observational astronomy was wrested from Europe and brought to the United States. If any one name could be attached to this achievement, it would be that of George Ellery Hale, one of history's great scientific entrepreneurs. Following the example pioneered at Lick Observatory, Hale chose another California peak, Mt. Wilson, as the site for his new observatory, and equipped it with two great reflecting telescopes. Hale came to California from the Yerkes Observatory, where he had built the world's largest refracting telescope, using a lens 40 inches in diameter. Hale realized that it would not be practical to build larger lenses, so he turned to the more promising technology of glass mirrors. With the financial assistance of Andrew Carnegie, he first built a reflecting telescope with a 60-inch mirror (Figure 2.1), and then a 100-inch version. The optical surfaces were finished so finely and the mechanical systems were engineered so well, these two telescopes set standards of excellence and scientific productivity that prevailed for decades.

The construction of the Mt. Wilson Observatory marked the beginning of a golden age in American astronomy. In the 1920s,

Figure 2.1: *George Ellery Hale (right) and Andrew Carnegie standing in front of the new 60-inch telescope dome on Mt. Wilson. Note that Carnegie has had a minor problem with his coat buttons.* (Mt. Wilson Observatory photograph, courtesy Owen Gingerich)

there was Hubble's demonstration that the spiral nebulae are "island universes" much like our own Milky Way galaxy, followed in the 1930s by his discovery that the universe is expanding. These triumphs spurred the construction of the 200-inch reflector on Mt. Palomar (Figure 2.2), still the most productive telescope of this century, with important contributions in many fields of modern astronomy such as quasars and observational cosmology. The 200-inch was so successful, both scientifically and technically, that it became the model for a whole generation of modern telescopes—many of them overseas, several of them very successful, but none of them bigger. These facilities have both contributed to the blossoming of modern observational astronomy and served a bur-

Figure 2.2: *The 200-inch telescope on Palomar Mountain. This design set the standard for a generation of large telescopes.* (Palomar Observatory photograph)

geoning community of young scientists attracted to the field by the excitement of the Space Age.

It is striking that the revolution in modern ground-based optical astronomy has not been due to the construction of much larger telescopes. The workhorses of today are typically only 50 percent larger than the Mt. Wilson 100-inch telescope built almost 75 years ago. Rather, it has been the introduction of new instruments using modern photoelectric and solid-state detectors, combined with precision computer control and rapid data handling, that has brought about the revolution.

For example, consider the detectors used by astronomers to measure and record extremely faint sources of light. Twenty-five years ago, astronomers mainly used photographic plates. For many years, the best photographic materials for astronomy were manufactured by Eastman Kodak. This was the result of a corporate commitment by Kodak to work with astronomers on the development of materials optimized for astronomy. Many of the important developments, such as chemical treatments that improved the sensitivity of the plates for the very long exposure times typical in astronomy, or extended the spectral reponse from the classical blue region out into the yellow and red, were the work of C. E. K. Mees, a brilliant scientist in the Kodak Research Laboratories. Mees had a personal interest in astronomy, and he was no stranger at Mt. Wilson. (The late Bart Bok, one of astronomy's great raconteurs, liked to tell how Mees happened to come to Rochester in the first place. George Eastman wanted to develop the capability for making colored gelatin filters so that these could be added to his product line. He heard that there was a young chemist named Mees in England who had become expert in the making of such filters, and Eastman went to England to see if he could hire Mees away from Wratten and Wainright. Mees was interested in Eastman's offer, but turned it down because of the loyalty he felt to his company. This didn't stop George Eastman, however; he simply bought Wratten and Wainright.)

Also, 25 years ago, even the very best photographic materials used for astronomy wasted at least 99 percent of the light collected by the telescope. It is rather horrifying to think that out of every 100 photons of light reaching the telescope, less than one had a chance of being recorded. Moreover, the photographic process has a sensitivity threshold; if the amount of light does not exceed some minimum level, no image at all is produced.

A major advance over photography, at least in terms of the light-detecting efficiency, resulted from the growing use of an entirely different class of devices—ones utilizing photoemissive cathodes inside of vacuum tubes. Photomultipliers, a version of this vacuum-tube technology that incorporated amplification of the electronic signal, found widespread use in the 1950s and 1960s. (Photomultipliers were available long before this era, but practical and reliable units were not widely available until after the technological developments spurred by World War II.) Photomultipliers typically can record up to 10 or 20 percent of the incoming light, an improvement of more than a factor of 10 over photography. The major drawback of the photomultiplier is that it has just one channel and can only record the total amount of light in an image, without giving any details of the image structure. Since stars are point sources anyway, this was not a particular disadvantage for stellar observations, and photomultipliers became the standard detector for precise measurements of the amount of light coming from individual stars.

More general versions of photomultipliers, ones in which the details of the image were preserved and intensified, became common in astronomy by the 1970s. Although specialized image intensifiers had been used earlier, it was once again the intense technological development for military applications that made reliable and rugged devices widely available. The army's desire to view enemy vehicles and personnel in a Southeast Asian jungle by available starlight on a moonless night helped create image intensifiers that also became widely used in astronomy, both as the main detectors in observing instruments and for the television viewing systems that brought astronomers out of the cold dark night and into warm lighted rooms.

The most exciting modern development in visible-light detectors was the advent of silicon diode arrays with efficiencies as high as 70 or 80 percent. Silicon is such a sensitive detector of light that it usually wastes less than it uses. In terms of raw sensitivity, this

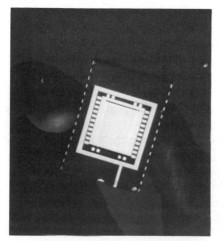

Figure 2.3: A CCD built by GEC. This large integrated circuit has an array of 385 × 576 diodes and a peak efficiency in excess of 40 percent for the detection of visible light. (Photograph Copyright 1985 by Ross Chapple)

means astronomers are now approaching the theoretical limit possible at visible wavelengths. On the other hand, the diode arrays have limited sizes. Charge-coupled devices (CCDs), the most common form of diode array used for direct imaging in astronomy, are typically not much bigger than 1 × 1 centimeter. (Figure 2.3.) The technology will surely evolve to allow larger formats, and at least one manufacturer is talking about CCDs that will measure 5 × 5 centimeters. This is still far short of the 50 × 50-centimeter photographic plates now used by the 100-inch du Pont telescope at Las Campanas, Chile.

A major advantage of diode arrays such as CCDs is that they can be interfaced directly with computers. The intensity of the light at each point in an image is easily available as a digital number that can then be fed into a computer where it can be processed and displayed using modern video technology. It is interesting how the development of more powerful computers is just managing to keep up with the ever-increasing size of the diode arrays. Present CCDs typically have less than a half-million individual picture elements (pixels), and their output can be handled with 16-bit computers. However, the next generation of solid-state imagers will require 8 megabytes of memory just to store one image, and this exceeds the address space of the older computer technology. The new 64-bit work stations are coming available just in time.

Although there will undoubtedly be important further developments in the technology of visible-light detectors, especially toward larger sizes, there can be no more spectacular gains in basic sensitivity, simply because silicon is already close to what is theoretically possible for a "perfect" detector. Perhaps this is why attention is now turning back to the construction of much larger telescopes, because one way to press on to fainter and more distant objects is to increase the collecting area of the telescope.

For observations of very faint point sources, another important way to improve the observing limit of a telescope is to find a way to get sharper images. For ground-based observatories, density variations in the atmosphere above the telescope distort the incoming image so that a point source jumps around and gets blurred out, a phenomenon known as "seeing." The second part of the problem is that there is a general glow to the night sky—even when the Moon isn't up. For very faint sources, the glow can be much brighter than the source itself. Thus the problem is to extract the signal of the source out of the surrounding and unwanted glow. If a point source image is blurred out by the seeing to cover a larger image area, then it is even harder to extract its image from the general glow. If the seeing is better, however, then the point source produces a sharper image, with stronger contrast above the surrounding background. This is why the Hubble Space Telescope will have such an advantage over most ground-based telescopes. In space, the image size of faint point sources will be limited only by the quality of the telescope optics, hopefully to a size of less than 0.1 arcsecond. When compared to most ground-based observatories, where the seeing is typically 1 or 2 arcseconds at best, the Space Telescope has a big advantage.

At the best ground sites, however, the seeing often can be better than 1 arcsecond. At Mauna Kea, for example, images as small as 0.3 arcsecond are not uncommon. Because the advantage gained by sharper images goes as the square of the image size, the quality of the astronomical seeing is just as important as the diameter of the telescope for research on faint point sources. Thus it is vital to find observatory sites that have the right conditions for good seeing.

For some kinds of research, the seeing is not so critical. Galaxies have fuzzy images that never get much smaller than 1 arcsecond in diameter. Thus, for research on faint galaxies, only a modest advantage is gained by going into space. And this advantage can be more than compensated for on the ground by having a much larger primary mirror—or by having much smaller construction costs. This is one reason there is so much interest in building larger telescopes on the ground. However, there is an even more fundamental and more philosophical reason for creating a new generation of ground-based telescopes. The powerful new orbiting observatories, such as the Hubble Space Telescope or the Advanced X-Ray Astronomy Facility, won't usurp the astronomy being done from the ground; rather, they will make new discoveries and open new areas of research requiring extensive follow-up and complementary observations.

For example, it is typical to say that an X-ray source has not been identified until an optical spectrum has been obtained for all the possible candidates that show up as optical images near the X-ray position. This is because an optical spectrum can tell so much about the physical characteristics of the source. If the spectrum has the absorption lines characteristic of a star, the spectrum can be used to determine what kind of a star is involved. If there are a few broad emission lines, then the source is likely to be a quasar, and its distance can be estimated from the redshifts of the emission lines. If there are sharp emission lines, but still with a substantial redshift, then the object is almost certainly a galaxy that has an active nucleus. In general, for any extragalactic object affected by the expansion of the universe, the redshift measured from an optical spectrum immediately gives a reasonably accurate distance, which, in turn, allows the intrinsic luminosity of the source to be calculated.

Thus, in the 1980s, a new generation of giant telescopes is being conceived, designed, and, in some cases, funded and built.

The first step toward this new generation—and away from the conventional approach as defined by the Palomar 200-inch—was taken by the Soviets. In the 1960s, they set out to build the world's largest optical telescope. (Figure 2.4.) For the tracking mount,

Figure 2.4: *The Soviet large astronomical telescope. The 6-meter primary mirror is carried in an altitude-over-azimuth mount. The observatory is located in the Caucasus Mountains, near the scientific village of Zelenchukskaya.* (Photograph by David Latham)

they chose an altitude-over-azimuth (alt–az) arrangement, with one axis pointing at the zenith instead of at the north celestial pole, as is the case for the classical equatorial mount. This decision committed the Soviets to the development of the sophisticated direction encoders and computing systems needed to control simultaneously the varying speeds of the two axes with the high precision demanded for optical observations. The Soviet 6-meter (236-inch) telescope is now in routine operation, and all future giant telescopes will almost certainly use "alt–az" mounts, because of the advantages gained from only having to deal with gravitational forces that change around one axis instead of two. These advantages more than compensate for the high technology required to achieve accurate tracking.

The difficulty of fabricating and handling very large mirrors had long been viewed as the major stumbling block to the construction of single-mirror telescopes much larger than the 200-inch. The troubles that the Soviets encountered in making a good primary for their 6-meter monster certainly did little to dispel this opinion. One of the early attempts cracked spontaneously, and the first mirror that was actually installed in the telescope gave marginal images. Although the Soviets eventually achieved a primary with excellent optical characteristics, the mirror is solid glass and weighs about 50 tons. As a result, it takes a long time for this mirror to stabilize after the temperature in the dome changes, and this factor can degrade its astronomical images.

There are several ways to address the problem of thermal stabilization in the design of giant primary mirrors. One common approach is to use special glasses that have very low thermal expansion, such as Cervit or fused silica, so that the shape of the mirror is less affected by changes in temperature. However, this does not solve the problem that solid mirrors can store large amounts of heat, which must be dissipated as the temperature changes. The resulting waves of convection in the air near the mirror can disrupt the image sharpness. One solution is simply to locate the telescope at a site where the temperature never changes. This is not as impossible as it may sound. For example, the summit of Mauna Kea on the island of Hawaii is usually above the temperature inversion layer, and sometimes the temperature hardly changes for days at a time. This is undoubtedly one of the reasons that Mauna Kea is famous for its excellent image quality, despite the fact that the major telescopes located there have solid primaries.

Another approach to the thermal problem is to fabricate the primary mirrors using structures with thin sections, both to reduce the total heat stored in the mass of glass and also to make it easier for the glass to exchange its stored heat with the air. (Lightweight mirrors also have the advantage that they reduce the total mass that must be carried by the telescope mount.) One way to achieve thin sections is to build the mirrors with hollow structures; another is to use external structures, or active controls, either to ensure the shape of a single large thin mirror or to combine the light gathered from several smaller and lighter mirrors.

Both these approaches were used for the Multiple Mirror Telescope (MMT) located on Mt. Hopkins in Arizona. (Figure 2.5.) The MMT is actually six independent telescope systems carried on a common alt-az mount, with the images from the six telescopes brought together to a common focal plane. (Figure 2.6.) Each of the MMT's six primary mirrors was built from fused silica plates typically one inch thick. The front and back plates are 72 inches in diameter and are separated by an eggcrate type of structure assembled from interlocking pieces. Before grinding and polishing, the mirrors were heated up just to the exact point where the individual pieces fused together and the entire structure sagged into the approximate concave shape needed. (A few more degrees of heat and the mirror blank could have sagged into a pool of liquid glass!) Obviously, this type of mirror construction is not cheap and probably would not have been chosen for the MMT had the mirror blanks not been available on surplus from a Department of Defense program to build spaceborne telescopes that would be pointed down instead of up.

To hold the images from the six telescopes together exactly, the MMT's designers originally intended that the pointing angle of the secondary mirrors of each telescope would be adjusted continuously, using a laser reference system. A phototype laser system was built and installed, but, in practice, it did not perform as well as a skilled operator, who watched the images in the main focal plane on an intensified television viewer and adjusted the alignments of the secondaries by hand. As a result of this experience, the philosophy of how to keep the images aligned was changed. The laser system was removed from the telescope, and a computer-controlled system was developed for doing automatically what the

Figure 2.5: *The MMT, showing the six primary mirrors. This facility is a joint project of the Smithsonian Institution and the University of Arizona.* (Smithsonian Institution photograph by Dane Penland)

operators had been doing by hand. Since this system for coaligning the telescopes requires that there be a suitable point source image in the field of view, it gives away the possibility of accurate blind pointing and tracking. For example, it was originally hoped that the MMT would have absolute pointing to an accuracy of about one arcsecond, so that it could be used for infrared observations during the day even though no television images of stars would be available. Actually, the loss of this capability has not upset the optical astronomers, who always worry that exposure of their telescope to excessive heating during the day will degrade the image quality on the next night.

It never ceases to amaze me how many groups around the world are interested in building giant telescopes. In the United States

Figure 2.6: *The beam combiner for the MMT. The beams from the six individual telescopes are brought together by the six-sided prism. Normally an instrument such as a spectrograph would be mounted just below the beam combiner.* (Smithsonian Institution photograph by Dane Penland)

alone, at least a half-dozen serious design concepts for telescopes larger than 5 meters have been described in print. In addition, there are ambitious efforts underway in Europe, Japan, and the Soviet Union. For example, when I visited the Soviet 6-meter facility in 1981, I asked the director why they had built the telescope at the 6,000-foot level instead of pushing on up to the 9,000-foot level of the same mountain, which loomed invitingly in the background. I suspected the reason was that the project had run out of the money to build the road any farther, but the reply was that the summit area was being reserved for a future 25-meter monster! There is something about a giant telescope that has an almost universal appeal. Even the Saudi Arabians talk of building the largest telescope in the world.

Instead of cataloging all the giant telescope dreams, I will mention two efforts that seem to have the best chance of ultimate success: the California 10-meter telescope (TMT) and the Arizona "big-mirrors" project. These two projects have both identified the primary mirror(s) as the most fundamental problem to be solved, and each has come up with a rather different solution.

The Californians have adopted the approach of putting together an array of 36 small hexagonal mirrors, each one fabricated to be a segment of the overall parabolic shape required for the assembled array. In this approach, two key technological problems must be solved: how to generate the correct surface shape for each mirror segment, and how to keep all the mirrors aligned so they form a common image.

To get the proper off-axis parabolic shape for the mirror segment, each mirror is bent by carefully calculated amounts while it is being ground and polished to a spherical shape. When the finished mirror is released, it springs into the required shape. This process is undergoing development, and has shown partial success in early tests. If the finished mirror segments do not spring into exactly the desired shape, the fallback solution is to apply small bending forces in the final mounting cell in order to achieve just the right shape.

The simultaneous alignment of all the mirror segments will be accomplished using an array of sensors located at the joints of the various mirrors. Each sensor can detect an extremely small local misalignment between its two adjacent mirrors. The signals from all the sensors are combined together by a computer system that then directs how much each mirror must be adjusted by actuators in its mounting cell. This process will go on continuously at a speed of many cycles per second.

The Arizona approach to solving the giant mirror problem is quite different, and largely results from the vision of Roger Angel, an astronomer at the Steward Observatory. In May 1979, at the banquet celebrating the dedication of the MMT, I sat at the table with Roger. The conversation naturally turned to what might be the next step, now that the MMT was officially in operation. Roger announced he was going to start work on the technology required to make gigantic, lightweight, and inexpensive mirrors, because this would be the stumbling block to be overcome if astronomers were to achieve the next generation of telescopes. I was rather appalled by the thought of starting out on a fundamental research program that might not reach its ultimate goal for 10 or 20 years. Instead, I was anxious to start using the telescope that we had just spent 10 years building.

Roger's vision proved to be much more farsighted than mine, and in a few short years he has made remarkable progress toward developing the technology needed to make a lightweight mirror 8 meters in diameter. Roger has concentrated on the techniques for casting large honeycombed mirrors, using inexpensive borosilicate glasses similar to the special Pyrex used for the 200-inch blank. Although the thermal expansion of this glass is not as small as fused silica or Cervit, Roger's concept is that the temperature of the entire blank can be brought to the ambient conditions rather quickly by forcing copious amounts of air throughout its hollow structure. The key is that none of the glass sections are very thick.

To keep down the cost of the telescope mount and its protective building, the focal length of the primary mirror must be as short as possible. In turn, this implies that the surface of the primary must be deeply concave. For such sharply curved parabolas, the deviation from a sphere is very large. This is a crucial issue, because the natural result of normal grinding procedures is a sphere. Several innovative features of the Arizona effort are aimed at producing the necessary deep parabola on an 8-meter blank. First, the entire casting oven will be spun rapidly on a turntable (Figure 2.7) while the glass is in its molten state, which leads naturally to a rough parabola. The shaping of the mirror to its final parabola will be carried out on a large optical grinder, the largest of its type in the world (Figure 2.8), which has already been installed and used successfully on a smaller, less extremely concave mirror. In the final polishing, the lap will be bent into the necessary parabolic shape, with the actual shape changing as the lap moves to different positions on the mirror.

The California TMT project received an enormous boost in Jan-

Figure 2.7:   *The oven for casting large mirrors at the Steward Observatory of the University of Arizona. The oven is mounted on a turntable, which spins at a constant rate while the glass is molten in order to create the rough parabolic shape of the mirror blank.* (University of Arizona photograph)

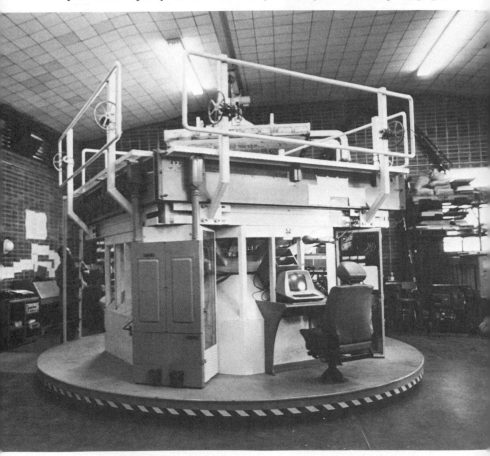

Figure 2.8: *The large optical generator at the Optical Sciences Center of the University of Arizona. This is the largest precision grinder in the world, capable of handling mirrors as large as 8 meters in diameter.* (University of Arizona photograph)

uary 1985 with the announcement of an intended gift of $70 million from the William Keck Foundation to the California Institute of Technology (Caltech). The early development work for the TMT was carried out by scientists at the University of California (UC), and they will continue to be responsible for the design. With the

Figure 2.9: *A model of the California 10-meter telescope.* (California Institute of Technology photograph)

Figure 2.10: *A model of the National New Technology Telescope. The beams from the four independent 8-meter telescope systems can be brought to a common focus, or each of the telescopes can be used with its own instruments for simultaneous complementary observations in the same general field of view.* (National Optical Astronomy Observatories photograph)

joining of forces between Caltech and UC, the funding for this project seems secured—and almost entirely from private sources.

The TMT (Figure 2.9) will be located at Mauna Kea, which is fast becoming the world's premier ground-based observatory site, with several major instruments already located there. The Hawaiians like to say their observatory is "halfway to space," because the good seeing on Mauna Kea already realizes much of the advantage it is expected the Hubble Space Telescope will have. In addition, because of its extremely high altitude, Mauna Kea is above much of the water vapor that interferes with infrared and submillimeter observations.

The Arizona big-mirrors project received support in 1984 with the announcement that the National New Technology Telescope (NNTT) would use the multiple-mirror concept and would be built with four 8-meter mirrors. (Figure 2.10.) Although the biggest mirror that Roger Angel and his crew have cast so far is only 1.8 meters, plans are laid for progressing in steps to the 8-meter size over the next few years. One obvious step preceding the "four-shooter NNTT" would be a single-mirror 8-meter telescope. This would both allow evaluation of mirror performance and encourage the development of focal-plane instruments that could immediately work on the NNTT. As yet, the site for the NNTT has not been

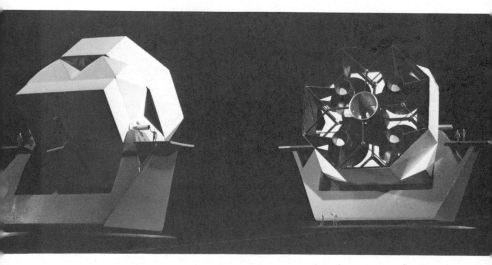

chosen. Two candidate mountains now being compared carefully for their astronomical qualities—especially the seeing—are Mauna Kea, and Mt. Graham in southeastern Arizona.

These are but two of the efforts to build a telescope larger than the Soviet 6-meter. Indeed, the race is on to build the world's largest telescope, and the decade of the 1990s may see this title change hands two or three times.

Speckle Imaging

Astronomical Image Recovery by Innovative Techniques

ROBERT STACHNIK

During its millions of years of travel to Earth, the light of a star is remarkably unaffected by its passage through the interstellar medium. Nonetheless, on this journey—from raging stellar surface to eye of a child, typically a distance of 300,000,000,000,000 kilometers—starlight acquires the properties that make it "twinkle." Depending upon one's point of view, it is either charming or horrible that this occurs as it passes through the Earth's atmosphere during the last 10 kilometers of the trip.

Our atmosphere is often described as an "ocean of air." The analogy brings to mind a phenomenon familiar to anyone who has ever spent hours watching kids play in a large swimming pool on a sunny day—the slightly mesmerizing pattern of bright lines that play across the bottom. Seen from the bottom of the pool looking up, the Sun appears to brighten and darken as bright or dark areas cross one's eye. The illumination change occurs as a result of a crude focusing process associated with the turbulent water surface. If the pool's surface was perfectly smooth, the effect would disappear.

Turbulence also exists in the Earth's atmosphere and produces similar focusing phenomena. One effect is twinkling. The romantic observer may feel this atmospheric turbulence creates a thing of beauty. However, for the optical astronomer, the annoying consequence is image degradation. For much of the last century, astronomers have tried to repair—or, at least, overcome—the damage done to their images by the atmosphere. Most of the time, they have simply accepted the problem. Now, innovative techniques of recovery and processing, *speckle imaging*, promise new and improved ways of seeing the skies.

What does the image of a star look like through a very large telescope? Remarkably, the astronomical community did not have a good answer to that question prior to 1970, some 125 years after William Parsons, the third Earl of Rosse, began observing with a 72-inch aperture reflecting telescope, dubbed "The Leviathan."

For a small telescope, the answer to our question is an easy one. It was provided in 1871 by Sir George Airy, then Astronomer Royal of Great Britain. Telescopes of good optical quality having apertures less than 10 centimeters in diameter image stars essentially as they would point sources. Their ability to resolve fine spatial detail, that is, to separate two close stars or to distinguish features on the surface of the Sun or Moon, is set by a fundamental physical consideration: the wave nature of light. For an imaged point source, of arbitrarily small angular size, Airy determined that an optically perfect telescope would produce a patch of illumination whose characteristic size is a function of the telescope aperture and of the wavelength of light. The diameter of this patch, in arcseconds (*theta*), is given by the expression:

$$\text{theta} = 0.00244 \times \text{lambda} / d$$

where *lambda* is the wavelength of the light in angstroms and *d* is the telescope diameter in centimeters. Thus, *theta* is the "diffraction limit" of the telescope. For a 10-centimeter telescope and violet light (4,000 angstroms), the resolution would be about an arcsecond. (For reference, a dime seen at a distance of four kilometers subtends an angle of 1 arcsecond over the sky.)

The detailed appearance of this "patch of illumination" is well known—a bright central region surrounded by concentric circles. Figure 3.1 is a typical "Airy pattern." Again, the angular size of

Figure 3.1:
An "Airy pattern."

the source is much smaller than the diffraction pattern, which is purely a consequence of optical interference effects.

Airy's equation suggested that the resolution capability of the largest telescopes should be about one-fiftieth of an arcsecond. In fact, no matter how optically good the telescope, one is rarely able to resolve details finer than an arcsecond due to atmosphere-associated image degradation. Moreover, the degree of image degradation is highly variable and unpredictable; thus, some experiments on the biggest telescopes can be effectively shut down even on nights when the sky appears crystal transparent because the "seeing" has blown up to as much as 20 arcseconds.

Therefore, despite Airy's equation, the original question—what does the image of a star look like through a very large telescope?—remained unresolved for nearly a century. An answer was proposed in 1969 by a young French scientist who at the time had never, himself, looked through a large telescope. Antoine Labeyrie suggested that, rather than the almost randomly scrambled image one might expect, the image structure should be dominated by the same optical-interference effects that produce the complex array of tiny, bright features characteristic of those images made with laser light. These bright features are called "speckle," and they result from one of the laser's unique properties: its ability to produce "coherent" illumination.

The notion that waves, be they water, sound, or light, can interfere "constructively" or "destructively" is a common one. For example, two ocean waves may combine at some point offshore to produce a resultant wave that has a larger (or smaller) amplitude than either wave alone. Naturally, the character of the resultant wave depends on the waveforms and wavelengths of the two constituent waves and on their relative "phase." In the most simple demonstration of this effect, assume that the waves have the same shape and length. Figure 3.2 (a) and (b) illustrate the result of adding identical waves that are exactly in phase and waves that are one-half wavelength out of phase. Similarly, the wave in Figure 3.2 (c) can be added to itself after varying amounts of fractional displacement to produce a wide range of waveforms that nevertheless remain the same over many cycles. By contrast, this is not the case in Figure 3.2 (d). The waveform changes after some number of cycles. Consequently, it cannot be added to itself after displacement by an arbitrary number of wavelengths to give the same result.

Figure 3.2 (a) and (b) also illustrate one aspect of the concept of optical coherence. Light emitted by thermal sources (those that glow because they are hot) usually is not highly coherent. Unpredictable disruptions in the waveform appear frequently, a situation similar to Figure 3.2 (b). Light emitted by a highly coherent source, like a laser, suffers such disruptions only over many, many cycles, analogous to Figure 3.2 (a). In experiments utilizing laser light, various processes—scattering, reflection, to name two—may cause a surface to be illuminated by light that has traveled a variety of paths. The constructive and destructive interference seen as speckle is visible because the pathlength differences are short compared to the distance a ray of this highly coherent light travels before a cycle disruption occurs. This property is called *temporal coherence*. (Interestingly, the finest speckle one sees in a laboratory setting is the result of interference actually occurring on the retina of the eye. This "subjective speckle" is unique for each observer looking at a coherently illuminated surface.)

Labeyrie's background in optical physics enabled him to see an analog between a simple experiment in coherent optics and the image-forming process for a large telescope. If one shines a laser through a lens, an Airy pattern appears at the focus of the lens. If one places a "diffuser," such as a rough-surfaced "ground" glass,

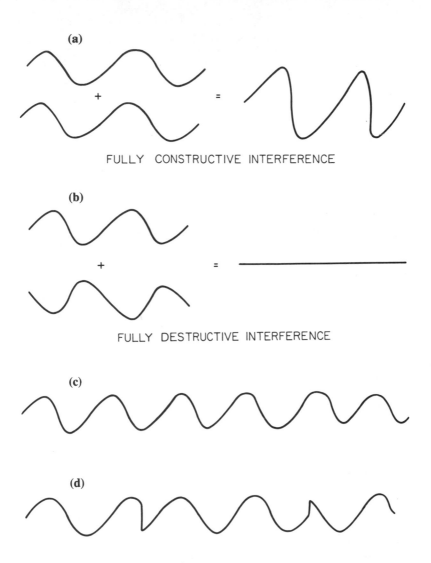

(a)

+   =

FULLY CONSTRUCTIVE INTERFERENCE

(b)

+   =

FULLY DESTRUCTIVE INTERFERENCE

(c)

(d)

Figure 3.2:   Wave interference: constructive and destructive. (Smithsonian Astrophysical Observatory illustration)

in front of the lens, the typical Airy disk is replaced by a large bright patch covered with speckle. If the laser beam illuminates the entire lens, the size of the speckle is set by the diffraction limit of the lens.

How is this similar to the image-formation process occurring at

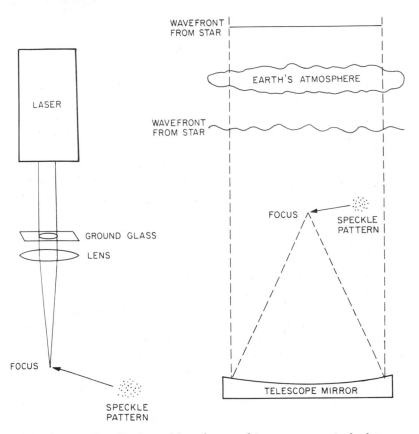

Figure 3.3: *Speckle formed by a laser and in an astronomical telescope.* (Smithsonian Astrophysical Observatory illustration by Joseph Singarella)

the telescope? Above the Earth's atmosphere, the wavefront due to a star is very nearly smooth and flat because of the great distance of the star. After passage through the atmosphere, however, the wavefront becomes corrugated, like light passing through ground glass. (Figure 3.3.) Portions of this corrugated wavefront become tilted away from the mean direction of propagation, resulting in illumination of a patch far larger than the Airy disk. The analogy becomes complete if it can be shown that there is a high degree of coherence for light striking different parts of the telescope's light-collecting surface. In fact, one of the founders of modern optical physics, Albert Michelson, showed precisely this in an important series of experiments early in this century. Michelson's "stellar interferometer" was a conceptually simple, but structurally challenging, instrument that collected light from a star at widely separated points along the wavefront, as far apart as 18 meters, and brought the beams together over optical paths that differed in length by microns. For all but a handful of "resolved" stars, the combined beams showed high contrast interference fringe patterns. Interference of light collected from widely separated points on a wavefront indicates a high degree of "spatial," as opposed to "temporal," coherence. As seen by the largest optical telescopes, such as the 6-meter Russian instrument in the Crimea, the light of most stars is highly spatial-coherent.

In the early 1970s, Daniel Gezari and I were graduate students working with Labeyrie at the State University of New York at Stony Brook. We had been granted a small amount of crepuscular (morning and evening twilight) observing time at the Mt. Palomar 5-meter telescope to photograph speckle. We planned to precede that experiment with a warm-up attempt using the Smithsonian's 1.5-meter telescope at the Whipple Observatory on Mt. Hopkins, Arizona. After nearly a month of partial nights at Mt. Hopkins— a considerable amount of observing time—we set off for Palomar without having once seen or photographed speckle. Antoine had claimed to have observed speckle on the spectrograph slit jaws at the 5-meter during an earlier visit, but, as second-year graduate students preparing to encounter the legendary Palomar telescope, we might be forgiven for feeling anxious.

Labeyrie was completely vindicated. The speckle looked better than the simulations we had done in the lab. In three partial nights, we not only recorded speckle on film but managed to make direct measurements of a number of supergiant stars for the first time since Michelson measured them, very laboriously, 50 years earlier. Figure 3.4 is a typical speckle image from that experiment. (We later learned that a stepper motor drive in the Mt. Hopkins telescope, since replaced, was the source of high-frequency vibrations that would have gone unnoticed by most other observers, but could well have smeared the speckle.)

Many useful astrophysical measurements can be derived from data such as Labeyrie, Gezari, and I obtained at Palomar, but it is worth describing first the special requirements for recording stellar speckle. These requirements are interesting because they

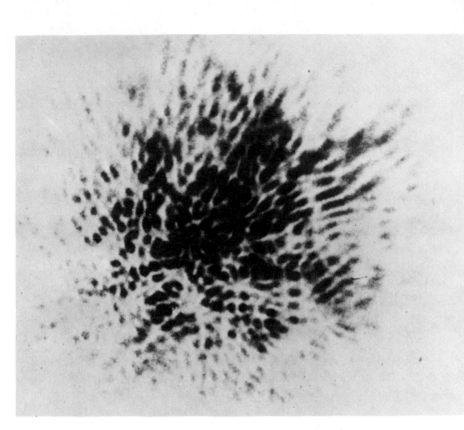

Figure 3.4: *Speckle image obtained at the Palomar telescope.*

both illustrate the instrumentation needs and illuminate the physics of the imaging process. The principal recording constraints are exposure time, magnification, spectral filtering, and atmospheric dispersion compensation.

Extremely short exposure times are required by the fact that the conditions of atmospheric distortion are constantly changing and recovery of useful data requires accurate recording of the instantaneous speckle pattern. Long exposures record only the 1 arcsecond (more or less) smearing of the stellar image. Because it is possible to see speckle in stellar images with the unaided eye, one assumes that speckle patterns usually do not change on time scales much faster than a few hundredths of a second. In fact, a typical exposure time is one hundredth of a second.

An important reason that the speckled character of stellar images was not widely recognized earlier is the fact that most astronomical experiments do not require very high magnifications of individual stars. In the case of a large telescope, the speckle may be 50 times smaller than the seeing disk. Astronomers normally have little reason to examine a distorted seeing disk in such great detail. In direct optical or photographic imaging, it is usual to match the finest structure one expects to see to the resolution of the detector, either the eye or the photographic plate, for instance. In such a case, one would never detect speckle. Several observers had actually seen speckle before Labeyrie, but none fullly understood its import or how it might be exploited.

Very high magnification is also needed—ideally, two detector resolution elements per speckle—and this has an interesting ramification. Not only does the painfully short exposure time limit our access to precious photons but we spread them very, very thin over the detector surface. After all, increasing the image scale by a factor of 50 means that the energy density per unit area decreases by a factor of 50 squared. The fastest available films were inadequate for our recording requirements. This necessitated development by Gezari and Labeyrie of an image-intensified movie camera.

The guts of the camera's film-advance mechanism came from an old projector purchased at a used-camera store on Thirty-third Street in New York City. The image intensifier was a then-new gizmo that converted an image formed on its front surface into a spatially modulated spray of electrons that were accelerated down the length of a vacuum tube and electrostatically focused onto a phosphor screen. The phosphor screen converted the image formed by the accelerated electrons into an optical image 100 times brighter than that formed on the front surface of the device. "Stacking" two of these tubes gave a total luminous gain of between 1,000 and 10,000, enough to observe stars as faint as ninth magnitude. Thankfully, such early homemade film systems have long since given way to modern electronic imaging sensors.

The third requirement for recording speckle is effective spectral filtering. The extraordinary temporal coherence of the laser is due, in part, to the spectral purity of the light it emits. Laser light is extemely narrowband. By contrast, starlight is emitted over a broad spectral range.

The effect of using broadband light in interference experiments is seen in Figure 3.5, in which the classical "Young's double-slit experiment" is performed with an angularly small thermal source, such as a pinhole illuminated by an incandescent lamp, and a nonnegligible spectral bandpass. Since the characteristic scale of an interference fringe pattern is a function of wavelength, the fringes at the red end of the bandpass will have a period different from those at the blue end. The scale varies linearly with wavelength, and white light fringes will be smeared after only a few fringes.

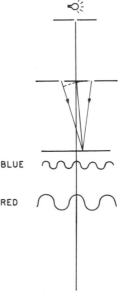

Figure 3.5: Young's double-slit experiment. (Smithsonian Astrophysical Observatory illustration by Joseph Singarella)

Speckle across a stellar seeing disk is smeared in the same fashion. At the Palomar telescope, for which there are 50 speckles across the seeing disk in 1-arcsecond seeing, we select a bandpass narrow enough to ensure that only at the outer edges of the image are the speckles smeared by a full speckle width. However, the narrow spectral bandpass (200 angstroms) means the number of photons reaching the detector is reduced again. Yet, even with a filter in place, radial smearing of the fringes/speckles can be seen in Figure 3.4.

For a star directly overhead, the foregoing requirements are all we have to meet. However, if the star is closer to the horizon, we must address the problem of atmospheric dispersion. When looking at any angle other than straight up through the Earth's atmosphere, we are actually looking through a weak prism. A hint of this effect can sometimes be seen when the setting Sun is very low in the sky: instantaneous atmospheric turbulence leads to the "green flash" seen just as the upper limb, or edge, of the Sun drops below the horizon. Atmospheric refraction near the horizon causes the Sun to appear almost its own angular width again higher than it actually is. This effect is slightly stronger in blue light than in red. Blue light is so strongly scattered by the atmosphere that a momentary atmospheric density fluctuation causing slight displacement of the red and blue limb images is perceived as a green, rather than blue, flash. At less extreme angles away from zenith, the effect for both Sun and stars is far less dramatic, and can be reduced still further by use of filters. Nevertheless, the speckles of stars are often seen to be smeared into little spectra, with their red ends toward the horizon. Some very ingenious optical designs have been developed to compensate for this atmospheric dispersion, one approach being the insertion of a pair of adjustable prisms to produce a variable compensating dispersion without causing deviation of the optical beam.

So far, we have a fairly good recipe for recording fine speckle structure within a stellar seeing disk, but have only hinted that this is in any way a useful thing to do.

For Labeyrie, the hard parts were recognizing that speckle even existed and then figuring out how to record the images. By comparison, the data analysis procedure was obvious—use of spatial power spectra derived from image Fourier transforms. (Perhaps not so simple, but at least understandable with a little explanation.)

The easiest way to see how high-resolution image information is coded into the speckle images is to look at Figure 3.6. Across the top are short-exposure direct speckle images of three stars taken at the 5-meter Palomar telescope. At the right is Alpha Lyrae (Vega), a star which is completely unresolved at the diffraction limit of the telescope. At the center is an image of Alpha Aurigae (Capella), long known to be a binary system from spectroscopic evidence and resolved by Michelson with his interferometer. Something is clearly peculiar about the speckle of Capella. This is evident even for direct telescopic inspection of the image by eye. The odd feature is that there are actually two identical speckle patterns

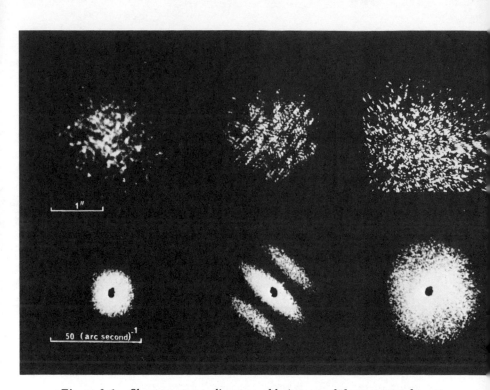

Figure 3.6: *Short-exposure direct-speckle images of three stars, taken at the 5-meter Palomar telescope: right, Alpha Lyrae (Vega); center, Alpha Aurigae (Capella); left, Alpha Orionis (Betelgeuse). Each of the three stars has below it a representation of the spatial power spectrum produced by an analog process. (The black spot at the center is an artifact of the technique.)*

here. Each speckle has a neighbor separated by the same distance, at the same position angle. Evidently, the wavefronts from both stars are passing through essentially the same chunk of atmosphere and are being distorted in the same fashion.

At the left is an image of Alpha Orionis (Betelgeuse). Again, the speckle looks different. The speckles are enlarged and overlap. This is because Betelgeuse is slightly resolved at the diffraction resolution limit of the telescope. The speckle produced by Vega measures a fiftieth of an arcsecond in size, as expected from diffraction theory. The speckle for Betelgeuse, however, measures about a twentieth of an arcsecond. This is a good approximation to the actual size of the star's image.

The formal description of the image-formation process includes the concept of "convolution." Every real optical system produces an image that is in some fashion a distortion of the object observed. One way to measure this distortion is to look at a single geometrical point; the resulting distribution of light around this point is called the "point response function." According to the Convolution Theorem, each point on the object is replaced by the instrument's point response function. Because the point response function of a large telescope has this speckled character, the degraded image will very nearly be an array of (overlapping) diffraction-limited images. As the object becomes significantly larger than the diffraction limit, as in the case of Betelgeuse, the amount of overlap increases and the apparent image contrast goes down. If we have 1,000 speckles, we will have 1,000 overlapped diffraction-limited images. For a still larger object, the overlap becomes more severe and the image becomes somewhat of a muddle. All the information required to recover a diffraction-limited image is, in fact, coded into individual short-exposure images, even if at reduced contrast.

The logical method for decoding these images is to make use of the Fourier transform, a mathematical operation that can represent an image in terms of the characteristic distances over which the intensity varies. The power spectrum, derived from the Fourier transform, tells one how much structure is contained in a given image at fine, coarse, and intermediate spatial scales.

Returning to Figure 3.6, each of the three stars has below it a representation of the spatial power spectrum produced by an analog process. (The black spot at the center is an artifact of the technique.) Comparison of the Betelgeuse and Vega power spectra shows them to be very different. For Betelgeuse (left), very little power is present in the outer parts of the spectrum, where the finest image structure is represented. The star is resolved, and both its diameter and information about its intensity profile can be deduced from this representation. Vega (right) shows far more fine structure (a larger power spectrum disk). The size and profile of the disk is just what one would expect for an unresolved point source. Capella (center) is different still; here the fringe pattern crossing the power spectrum is the unmistakable signature of a double star. From it, both the position angle and separation of the two bodies can easily be obtained. Fourier analysis is far easier than any process involving direct examination of the speckle images. Furthermore, many thousands of power spectra (from thousands of short exposure images) can be added together to produce resultant spectra showing extremely subtle features.

"Astronomical speckle interferometry," by which high spatial resolution information is recovered from short-exposure images, has become a mainstream astronomical technique. For example, it was the technique used to detect a companion object (planet or "brown dwarf") to the faint star Van Biesbroeck 8 in late 1984. The process, however, does not produce actual pictures.

The goal of producing images has attracted the attention of a number of scientists, and some interesting progress has been made following significantly different routes. Briefly discussing one approach may shed some light on the general subject.

The Fourier "transform" is simply an alternative way of representing image information. All the information contained in a speckle image is contained in its Fourier transform, and vice versa. From a computational standpoint, an image is just an array of numbers representing light intensity as a function of position in the image array. The Fourier transform of this speckle array becomes an array of the same size, but with an important difference. At each array location—each x and y coordinate—not one but two numbers are used to represent the function. This is another way of saying that the array is "complex," that an amplitude and phase are associated with each location.

The physical meaning of the amplitude and phase and how these quantities relate to the image can be best demonstrated by Figure 3.7, which shows a series of images produced by laboratory apparatus designed to simulate the effect of atmospheric turbulence on an astronomical image. The test object is a triple star. Figure 3.7 (a) shows how it might be seen in the absence of an atmosphere; Figure 3.7 (b) and (c) shows the associated amplitudes and phases. In fact, the amplitude array (more properly, the amplitudes squared) represents the result of conventional speckle interferometry processing. In other words, Figure 3.7 (b) is the power spectrum.

The problem with speckle interferometry processing is that there is no way to obtain the phase part of the Fourier transform of the high-resolution image. One can calculate it for, say, 1,000 short-exposure images, but unlike the amplitude-squared arrays, summing-phase arrays do not give the high-resolution Fourier transform information we want.

INPUT OBJECT

(AMPLITUDE)²

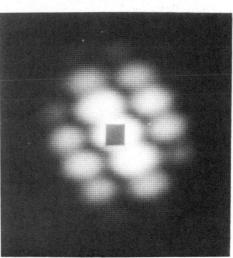

(a)

(b)

PHASE

(c)

Figure 3.7: Laboratory-simulated triple star image with computerized Fourier transform amplitudes and phases.

SHORT EXPOSURE INPUT FRAMES

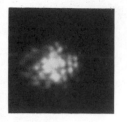

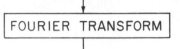

FOURIER TRANSFORM

AMP^2

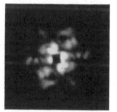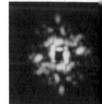

PHASE

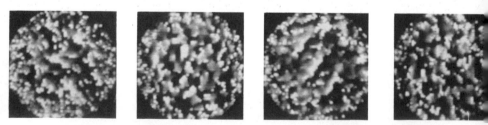

Keith Knox and Brian Thompson, at the University of Rochester Institute of Optics, proposed a solution later implemented by Peter Nisenson at the Harvard-Smithsonian Center for Astrophysics. Knox and Thompson proved that summing, not the phase arrays, but arrays of the *x*-direction and *y*-direction point-to-point phase

LONG EXPOSURE
IMAGE

AVERAGE AMP^2
4 FRAMES

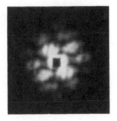

AVERAGE PHASE
4 FRAMES

Figure 3.8: *Speckle images produced by a laboratory atmospheric distortion simulator with associated Fourier transform amplitudes and phases.*

differences resulted in a quantity that could be translated into the high-resolution reconstructed image transform phase. This is at the heart of a process called speckle imaging.

Figure 3.8 illustrates speckle imaging using the lab simulator.

$(AMPLITUDE)^2$

SINGLE FRAME

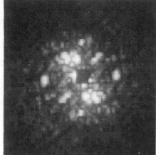

NO BIAS COMPENSATION

WITH BIAS COMPENSATION

Figure 3.9: *The results of summing arrays computed from Fourier transforms for many individual speckle images. The resulting recovered images reveal a triple star.*

Across the top are four instantaneous images of a mystery object. Summing the frames directly leads only to the conventional long-exposure image—very unenlightening. Also appearing in the figure are Fourier transform amplitude and phase arrays for each of the

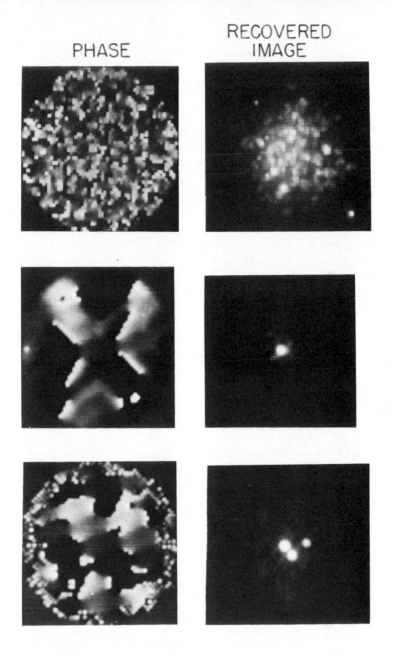

individual images. Again, not terribly interesting. They are too
noisy for any possible information content to show through. Summing even a small number of amplitude and differential phase
arrays starts to become interesting, however. The average ampli-

tudes and average phases seen at the right show hints of overall organization.

Figure 3.9 shows the result of summing a significant number of frames of data. Across the top are the amplitudes, phases, and recovered image for a single image frame. Across the bottom are the amplitudes, phases, and recovered image for 100 frames. The object is, of course, our triple star. (The center row shows the result of naive processing, in which no bias compensation for the effect of image noise due to photons is attempted.) Amazingly, the reconstruction at the bottom right of Figure 3.9 came from 100 speckle images just like those across the top of Figure 3.8 in which there is no visible hint of the triple structure. There is something faintly magical about that!

At this point, it is interesting to examine the cameras used to record speckle. Modern low-light-level cameras are a substantial subject in themselves. The most common sort of camera used in speckle applications is based on video technology, often used in conjunction with stacks of high-gain image intensifiers and photon-discrimination circuitry designed to suppress noise and ensure detection of all photons as pulses of identical height and shape. This latter feature is important for the photon noise-compensation procedure to which we have alluded. Many of these cameras are quite ingenious and have been very successful.

Another camera, different in design from video-based systems, has emerged from the work of Costas Papaliolios, Peter Nisenson, and Steven Ebstein at the Harvard-Smithsonian Center for Astrophysics. (Lawrence Mertz, at Lockheed Corporation, participated in early development.) This camera, the Precision Analog Photon Address (PAPA) detector, was developed specifically for astronomical speckle applications.

Figure 3.10 is a schematic diagram of a PAPA detector. The first element of the system is a very-high-gain image intensifier. The effective gain is of the order of 10^6, meaning that for every photon striking the front photocathode of the intensifier approximately a million photons are emitted at the back. An optical system capable of producing multiple images is placed behind the intensifier output screen. While many such systems can be devised, the one used here employs a single large lens to produce a collimated beam that then passes through an array of 19 lenslets. These small lenses, in turn, produce images of the intensifier output face

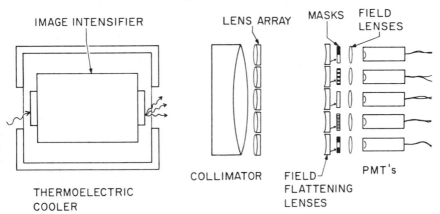

Figure 3.10: *Schematic design of a Precision Analog Photon Address (PAPA) detector.* (Smithsonian Astrophysical Observatory illustration by Joseph Singarella)

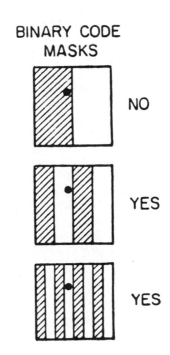

Figure 3.11: *Schematic of masking system in PAPA.* (Smithsonian Astrophysical Observatory illustration by Joseph Singarella)

on the surfaces of 19 small photomultiplier tubes. As described so far, we have a device that would, should a photon strike the front surface of the intensifier, produce 19 simultaneous signal pulses from the photomultipliers. The key to making this into an imaging instrument is a set of masks placed in front of the photomultipliers, at the foci of the lenslets.

The masks are transparencies that are partly opaque and partly light-transmitting. The principle of the masks is shown in Figure 3.11.

The first photomultiplier simply stares at the back of the intensifier, serving as a "strobe" channel, sending a signal whenever a photon is detected anywhere in the field. The second photomultiplier is covered by a mask, one half of which, say the left half, is opaque. When a photon is detected in the first channel, the

Figure 3.12: *The PAPA camera operated on a 2.1-meter telescope at Mauna Kea, Hawaii, by the author (right) and Costas Papaliolios.* (Smithsonian Astrophysical Observatory photograph)

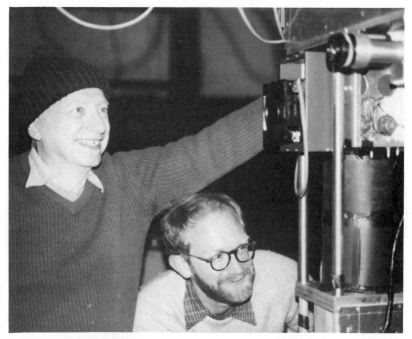

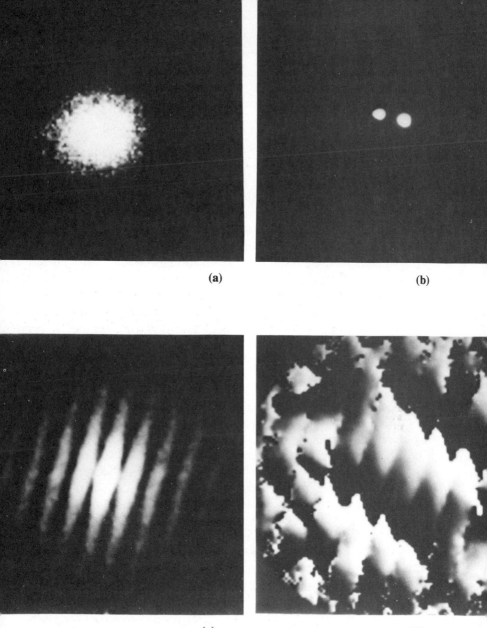

Figure 3.13: *Some examples of data processing from PAPA-recorded images: (a) direct sum of images, (b) reconstructed image, (c) Fourier transform (amplitude), and (d) Fourier transform (phase).*

second channel will detect it if it is on the right half of the field, but not if it is on the left half. We now know whether the photon is on the left or the right.

The third photomultiplier looks through a mask that divides each half, again, into opaque and transparent subareas. If the photon fell on the right half, we now know whether it is on the left or right half of the right half. The pattern from here on becomes fairly obvious. Successive masks have progressively finer patterns, and we have a binary-coded output from the photomultipliers that requires the addition of only a single optical channel plus photomultiplier to double resolution in one dimension. If we have n optical channels, the resolution in that dimension is 2^n. With nine channels, we have 512-element resolution in one dimension. To achieve the same 512-element resolution in the other dimension simply means adding another nine channels with masks rotated 90 degrees.

Figure 3.12 is a photograph of the camera mounted on the 2.1-meter University of Hawaii telescope at Mauna Kea; Figure 3.13 shows examples of the images and spatial power spectra the device is capable of producing.

The camera is unique because, without need for any analog-to-digital conversion, it provides the photon coordinates as a natural by-product of its mode of operation. More important, it provides the photon positions in a sequential catalog format, with times of arrival if desired. The photon catalog format is very useful at low light levels. Low-light-level imaging with videolike schemes usually means reading out (and possibly storing) lots of zeros. The catalog format is also useful when, as with speckle, exposure time is important because individual speckle frames can be reconstructed with as many or as few photons per frame as one desires. This amounts to changing the exposure time after the fact of recording.

Other devices that have contributed to producing high-resolution imagery include active optical systems. These devices are technological tours de force which, first, sense the atmosphere-induced deformations in an incoming stellar wavefront and then use a small mirror in the telescope's optical system to correct those deformations on millisecond time scales. A typical system of this type uses a Hartmann wavefront deformation sensor, shown schematically in Figure 3.14.

With the Hartmann sensor, starlight passes through the lens

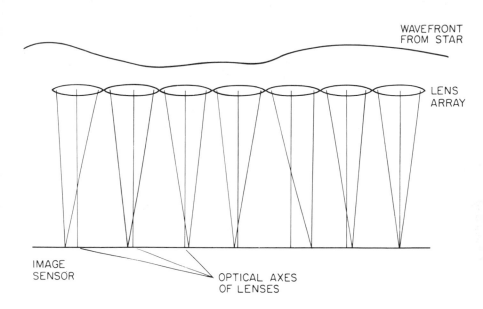

Figure 3.14: *A schematic diagram of the principle of wavefront distortion measurement using a Hartmann sensor.* (Smithsonian Astrophysical Observatory illustration by Joseph Singarella)

array, and the small local tilts in the wavefront introduced by the Earth's atmosphere are translated into displacements of the resulting images. If these displacements are continuously monitored, it is possible to calculate the shape of the wavefront deformation and send correction signals to the deformable mirror. The deformable mirror may consist of piezoelectric elements embedded in a somewhat elastic structure. The piezo elements are individually addressed and expand or contract as required to compensate for atmospheric effects. The problem is that the image displacements are exceedingly small (fractions of an arcsecond) and they occur exceedingly fast (in milliseconds).

Remarkable things have been done with active optical systems, but the devices suffer sensitivity restrictions that usually limit their use, in the "full atmospheric correction" mode, to brighter objects, or to simple "seeing" improvement. In fact, it is possible to use all the photons for the measurement, and have nothing left to bounce off the deformable mirror.

Much active optics technology has been developed in connection with laser fusion experiments, in which high-power-density laser beams must be precisely focused despite the thermal deformation of various optical components. Military programs also require active optics technology. Many groups are active in laboratory and theoretical investigations of high-resolution image-recovery procedures, but observationally oriented groups are most prominent in Britain, France, Germany, the Soviet Union, and the United States.

In solar system studies, speckle techniques have been used to improve images of the solar surface and to image both asteroids and the outer planets and their satellites, including Pluto and its moon, Charon. Numerous stellar diameters have been measured as functions of time and wavelength.

In other stellar studies, asymmetrical structures have been observed on some stars, and high-resolution polarization measurements have been made on others. (One star, Omicron Ceti [Mira] was seen to vary in size by a factor of 2 when observed at wavelengths differing by only 200 angstroms.) Recently, Harold McAlister, of Georgia State University, compiled a catalog of 3,363 interferometric measurements of 824 binary stars (the vast majority made by him) covering the dozen years since the introduction of speckle techniques. For example, Van Briesbroeck 8 is a binary, discovered with infrared speckle, whose second component is probably a planet-like brown dwarf.

Measurements have also been made for energetic galactic nuclei and so-called multiple quasars. Moreover, a wide range of important observations of galactic and extragalactic objects have been made in the infrared, and much has been learned about propagation characteristics of the Earth's atmosphere and about the theory of recovery of degraded images.

Certainly, the future of astronomical observation lies in space, but historic and powerful instruments, like the Palomar 5-meter, as well as a whole class of new-technology telescopes now being designed, will always remain on the ground. It seems clear that there is plenty of energy and ingenuity to keep them competitive for a long time to come.

Slicing the Sky

Sharper Images with an Orbiting Array of Optical Telescopes

WESLEY A. TRAUB

The first telescope appeared in Germany in 1608, and within less than two years Galileo Galilei had turned this new and exciting instrument toward the sky to discover mountains on the Moon, satellites around Jupiter, and a Milky Way that was not a glowing cloud but a mass of individual stars.

Galileo's observations were published in *The Starry Messenger* and his discoveries became widely known in Europe. These discoveries had an enormous impact in persuading people that the Earth should be viewed as being merely one object out of many in the universe, rather than as a unique object placed at the center. Indeed, it might even be argued that because of their stunning visual nature, Galileo's discoveries had a much greater effect than the theoretical arguments of Copernicus in the *Book of Revolutions* published 67 years earlier, or the grand summaries contained in the first and second laws of planetary motion published by Kepler in the *New Astronomy* only one year earlier.

What was this instrument—the telescope—that had such a revolutionary effect on then-current views of the universe? By Galileo's own account, his crude optical tube had an angular magnification about 33 times more than that of the unaided eye. (This is much more powerful than most modern binoculars, which have a magnification factor of only about 7.) Galileo's largest telescope lens had a diameter of 44mm (1.75 inches).

Let us ask, what was the increase in angular resolution compared with the unaided human eye? As we will see later, at nighttime our eyes have the capability of distinguishing two points of light if the points are separated by at least 120 arcseconds. This angle is one thirtieth of a degree, or about one fifteenth the size of the Moon. Aided by his 33-power telescope, Galileo's eye could resolve details that were 120/33, or 4 arcseconds, across. If the effects of geometrical aberrations in these early lenses were not too severe, then 4 arcseconds of angular resolution was certainly adequate to distinguish the brighter planets. For example, the disk of Saturn is 5 times larger than Galileo's resolution, Jupiter is 12 times larger, and at its maximum size Venus is 15 times larger from cusp to cusp. Let us also ask, how much fainter an object could be seen? As we will show later, at night the human eye has a pupil diameter of about 6.6mm. This means that Galileo's telescope had a collecting area which was about $(44/6.6)^2$, or 44 times greater than the unaided eye. With this increase in collecting area, the satellites of Jupiter would have appeared some 4 magnitudes more luminous, making them easily visible. These gains in angular resolution and collecting area were critical factors in making possible Galileo's discoveries—and for touching off the modern era of astronomy. It is quite conceivable that we could make similar gains, with respect to current observing abilities, within the next two decades, through an interferometric array of telescopes in space called the Coherent Optical System of Modular Imaging Collectors (COSMIC).

The Ideal Telescope

The potential gains—and limitations—of all telescopes, from Galileo's small refractor to the orbiting array of mirrors planned for COSMIC, are dependent on the diffraction and interference of light. These effects can best be demonstrated in a simple "ideal" telescope such as the one shown in Figure 4.1. The telescope is pointed toward a very distant star, so that arriving rays of light

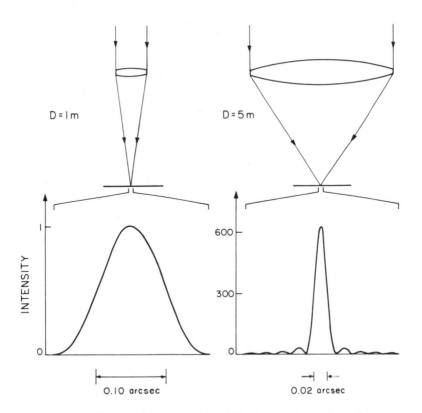

Figure 4.1: *The angular size of the diffraction pattern of a telescope is directly proportional to the wavelength λ and inversely proportional to the telescope diameter D. Here we show typical diffraction patterns for green light, with a wavelength of λ = 0.5 micrometer. It is interesting to see that in the case shown, the brightness of the center of the diffraction pattern increases by 625 times, where the diameter of the telescope increases by a factor of just 5. This large gain is the result of a factor of 25 from increasing the collecting area, and another factor of 25 from concentrating the light into a smaller spot size.* (Smithsonian Astrophysical Observatory illustration by Joseph Singarella)

are all parallel. The geometric action of the ideal telescope, whether it be a reflector or refractor, is to bring these parallel rays to a focus at a single point. Assuming that the lens or mirror has been polished with sufficient care to eliminate all geometrical aberrations, the telescope will then be limited only by the effects of diffraction. However, as shown in Figure 4.1, the parallel rays do not focus to a tiny point as might be expected; rather they spread out over an area having a characteristic size of about $f\lambda/D$, where f is the focal length, λ is the wavelength, and D is the telescope diameter. This means that if there are two bright stars in the sky separated by an angle greater than λ/D (radians), they will appear as separate spots of light. If they are closer together than about λ/D, the two stars will coalesce into a single blob of light. This angular resolution of about λ/D, or the wavelength divided by the diameter of the telescope, represents the *diffraction limit* of the telescope.* (The exact value of the angular resolution will depend on other factors, such as the shape of the aperture or if there is an obstruction such as a secondary mirror in the parallel beam.)

The Human Eye

Our eyes are also basically small refracting telescopes, with the lens diameter being set by the iris. In Figure 4.2 (b), the iris is shown closed down to a small diameter under bright light conditions, and opened up to a large diameter under dark conditions. What is the angular resolution of the eye at these two extremes? Using a thin-line source of light, the measured illumination of the retina in the human eye is shown by the solid curve in Figure 4.2 (c) for the two extreme cases of a small iris (diameter = 1.5 mm) and a large iris (diameter = 6.6 mm). The dotted curve is the theoretical diffraction-limited image of the same source. The amaz-

---

* Diffraction occurs because of the wave nature of light; but it can also be described as a result of the Heisenberg uncertainty principle, which tells us that we cannot simultaneously and precisely determine both the position and momentum of a photon. In a telescope system, a photon of light from a distant star must pass through the aperture, that is, the lens or mirror opening, which has a given diameter, D. The uncertainty principle says that if we know the photon's lateral position to an accuracy of about D, then its lateral momentum cannot be determined to better than approximately h/D, where h is Planck's constant. But the photon also has a forward momentum of h/λ, and the resulting angular deflection is the ratio of these quantities, that is, angular deflection = $(h/D)/(h/\lambda) = \lambda/D$. Thus, the incident photons are deflected by an angle that is typically of size λ/D, or approximately the diffraction limit of the telescope.

ing result is that under bright conditions (i.e., a small iris) the measured image comes very close to the theoretically predicted diffraction pattern, at least over the central part of the image. (Far from the center, there is a "skirt" of light caused by scatter and defects of focus.) On the other hand, under low-light-level conditions (i.e., a large iris) the measured image is nearly 7 times blurrier than predicted by diffraction theory. (This extra broadening may result from geometric aberrations that increase toward the edge of the lens, an effect commonly seen in simple lenses and mirrors.) The best angular resolution of the eye, then, is about 84 arcseconds in daylight, but only 120 arcseconds at night. In short, all planets are hopelessly unresolved by the human eye, since the maximum angular diameters for Venus and Jupiter are 61 and 47 arcseconds, respectively, with the other planets subtending even smaller angles. Small wonder then that the invention of the telescope had a major impact on astronomy.

Achieving High Angular Resolution

High angular resolution has been sought ever since Galileo's time, simply because it is an advantage in almost all astronomical measurements. (By common consent, astronomers use the word "high" to mean a *small* diffraction angle, λ/D.) And there are two important situations in which high angular resolution is critical. The first might be termed the "confusion-limited" situation, where we wish to measure an object in isolation from other nearby objects that might confuse the measurement if all objects were blurred together. There are many examples, including star counts in globular cluster cores, isolation of individual stars in the Milky Way and nearby galaxies, searching for identifiable features on the surfaces of nearby stars, measuring the sulphur volcanoes on Io, or measuring the shape of an asteroid. The second situation occurs when the object is faint and the zodiacal light background threatens to overwhelm the signal from the object. This is the case for objects or features that are at the resolution limit of the telescope, or smaller, because here a sharply defined beam pattern on the sky takes in the entire object, but excludes as much as possible of the smoothly distributed zodiacal light background. For very faint, small objects it is the random fluctuations in the number of counted

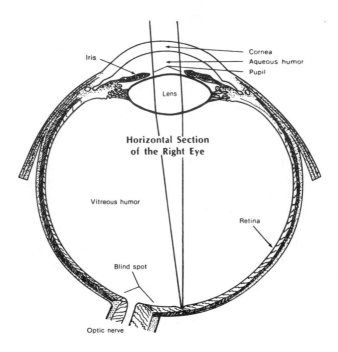

Cornea
Aqueous humor
Pupil

Iris

Lens

**Horizontal Section
of the Right Eye**

Vitreous humor

Retina

Blind spot

Optic nerve

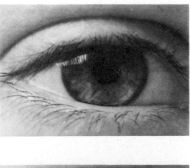

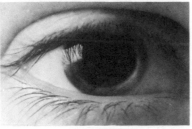

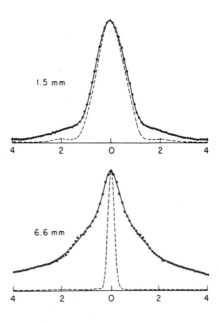

photons in the background that limit our ability to detect the object. Examples include all faint stars, many distant quasar nuclei, and many gravitational lens images.

The only situation in which high angular resolution is not an advantage is in measuring a featureless, extended object. In fact, it may be a disadvantage. Examples are relatively rare, since there are few truly featureless extended objects in the sky, but possible cases include extended sheets of excited gas from old supernovas, the interior regions of some planetary nebula halos, and the smoothly varying portions of elliptical galaxies.

The second important property of an ideal telescope is its collecting area, which is D^2, or $(\pi/4)D^2$, depending upon whether we are using a square or circular aperture. Intuitively, it seems that measuring capability should improve as the collecting area of the telescope increases. And, indeed, detailed calculations show that the signal-to-noise ratio of a measurement always increases in proportion to the square root of the collecting area.

Sharper and brighter pictures are therefore dependent upon the angular resolution (sharpness) and collecting area (brightness) of the telescope. The simplistic solution to the astronomer's quest is to build large-diameter telescopes. Alas, this approach quickly runs into problems, such as the strength of materials and budgetary limitations. In this chapter, we will see how astronomers can work around the problems by, for example, building telescopes with partially filled apertures, devices known as interferometers.

The first person to recognize that it would be possible to achieve high angular resolution without building prohibitively large diameter telescopes was the French physicist A. H. L. Fizeau. In 1868, he specifically pointed out the possibility of measuring the

Figure 4.2: (a) *Horizontal section of the right eye, showing ray paths of a distant point source of light. (b) The pupil is shown contracted and expanded for bright and dark ambient conditions, respectively. (c)* **The** *solid curve is the angular variation of the intensity of a line source as imaged on the retina for contracted (1.5 millimeters) and expanded (6.6 millimeters) pupils; the dotted curve is the theoretical image formed by a diffraction-limited lens having the corresponding diameter.* (Adapted from papers by Keith P. Bowen in *Sky and Telescope,* April 1984, pp. 321–324 [(a) and (b)]; and F. W. Campbell and R. W. Gubish in *Journal of Physiology,* Vol. 186, 1966, pp. 558–578 [(c)].)

angular diameters of astronomical objects. This suggestion was picked up by the American physicist A. A. Michelson in 1890 and, after nearly three decades of effort, culminated in his measurement (with Pease) of the angular diameter of Alpha Orionis (Betelgeuse), for which a value of 0.047 arcsecond was found. How did Michelson measure this diameter in the presence of atmospheric blurring, which is typically about 1 arcsecond?

Paradoxically, Michelson's technique was to cover over nearly the entire 250-centimeter-diameter mirror of his telescope, and use only two patches, each about 15 centimeters in diameter. The resulting diffraction pattern from either of the individual 15-centimeter subapertures was a relatively large blob in the focal plane, nearly 1 arcsecond in diameter. But, when both subapertures were used and the beams combined, this smooth blob turned into a corrugated pattern of straight bright and dark zones called *fringes*.

These fringes result from interference between the waves arriving alternately in phase and out of phase from the two apertures, since the difference in paths traveled by the two light beams changes by exactly one wavelength between adjacent bright (or dark) fringes. A cross-sectional view of the interference fringe pattern formed by the combined beams is shown in Figure 4.3 (c). The corresponding diffraction pattern of a fully filled aperture is shown in Figure 4.3 (a). If several small apertures are used to produce an array that has the maximum number of different distances between all possible combinations of pairs of mirrors (a *minimum-redundancy array*), the diffraction pattern will show a strong central spike and relatively small fringes, as in Figure 4.3 (b). (The patterns shown are for a single wavelength; if longer and shorter wavelengths are also included, then additional fringes will be superposed with wider and narrower spacings, respectively. These extra fringes will cause the dark regions between fringes to fill in, thus reducing the intensity contrast, or "visibility." If a very wide-band filter is used or, in the extreme case, no filter, then the one fringe that is formed with zero relative path difference will appear white and the adjacent fringes will be tinged with color.)

Why go to all this trouble? There is a very good reason: if we look through the atmosphere at a star using a small telescope, the diffraction pattern appears as a quivering, single blob. However, if we use a large-diameter telescope, the diffraction pattern breaks up into a rapidly changing swarm of points of light, *speckle*. (The

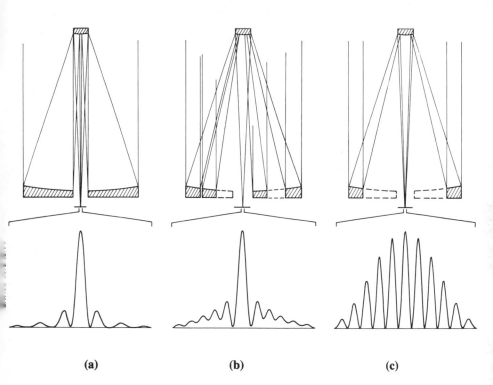

Figure 4.3: *(a) A single-telescope, one-dimensional primary mirror with a small secondary mirror, and the corresponding diffraction pattern, enlarged to show detail. (b) The same mirror with segments removed or covered so as to leave a particular pattern of equal-sized elements is known as a minimum-redundancy array. The diffraction pattern shows a strong central spike and relatively small sidelobes. (c) The extreme case of a nearly fully blocked central section, which leaves two widely separated mirrors. The diffraction pattern has many fringes spread over a relatively large angle, but the width of the central fringe is still quite small.* (Smithsonian Astrophysical Observatory illustration by Joseph Singarella)

boundary between *small* and *large* telescope diameters depends upon the degree of turbulence in the atmosphere, but it is generally taken to be about 10 centimeters.) This effect is shown schematically in Figure 4.4 (a), where the two wavefront segments are combined with no relative delay imposed by the atmosphere, so that the white-light fringe (with zero delay) forms in the center.

A short time later, typically a few hundredths of a second, the atmosphere will advance or delay one beam with respect to the other, so that at a given point in the focal plane the fringes will appear to shift sideways. The typical number of fringe changes is small if the distance between subapertures is small; large if the distance is great. For a separation of 500 centimeters (the diameter of the Palomar telescope), a typical change is about ± 5 wavelengths, or ± 5 fringes, as indicated in Figure 4.4 (b). (For additional details on "speckle," see the chapter titled "Speckle Imaging.")

Remarkably, these motions of the white-light fringe can often be tracked by a trained eye, which is what Michelson did. In essence, Michelson reasoned that each point on the surface of Betelgeuse contributed its own fringe pattern that was slightly shifted from the others in the focal plane. Thus, at some critical separation of subapertures, the bright and dark fringe patterns would complement each other and the corrugation effect would completely disappear. This was the basis of his measuring technique; and his determination of the diameter of Betelgeuse in 1921 was such a dramatic astronomical "first," it won both scientific and popular acclaim, even being reported on the front page of *The New York Times*.

Recall, however, that the diameter of Michelson's subapertures was deliberately chosen to be relatively small, about 15 centimeters, since with a larger aperture the diffraction pattern would tend to break up into a number of "speckles," as discussed above. The maximum allowable aperture that can be used is limited by the strength of atmospheric turbulence: under low turbulence conditions (when a star image appears to be relatively sharp), larger apertures are allowable. This all means that Michelson's technique is limited to relatively bright stars. Unfortunately, this is true for all other ground-based high-resolution observations. Although some innovative attempts are being made to compensate for atmospheric turbulence (see chap. 3) only by going to space can we eliminate completely its image-degrading effects.

Separated Telescopes

The Michelson stellar interferometer is an extreme example of a one-dimensional, *dilute-aperture* telescope. As the name sug-

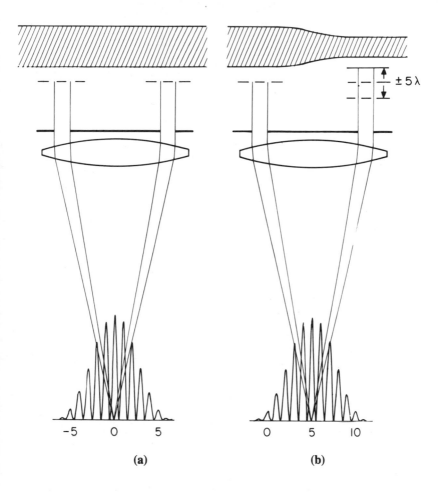

Figure 4.4: (a) Undisturbed (ideal) atmosphere produces a central fringe with zero relative delay between beams. (b) Disturbed (actual) atmosphere, with typical delay of about 5 wavelengths, produces a central fringe with a relative path difference of the same amount. (This delay is representative for an aperture separation of about 5 meters.) (Smithsonian Astrophysical Observatory illustration by Joseph Singarella)

gests, dilute, or *partially filled* aperture telescopes are those in which the primary light-collecting area is split into separated segments. The separation of the apertures, or receivers, allows distribution of a given amount of collecting area over a much larger-diameter region than would be possible if the same area was used as a single receiver. Since collecting area generally determines cost, this approach also can yield dramatic cost reductions with respect to conventional telescopes.

The Michelson stellar interferometer is considered *one-dimensional* because, when viewed from above, the mirrors are typically separated by a large amount compared to their diameters. This means that the high angular resolution direction is parallel to the mirror separation axis, and the low angular resolution direction is perpendicular. A one-dimensional instrument can be used directly on objects that are rotationally symmetric, such as slowly rotating, featureless stars. For more complex sources, either we need to make separate measurements along different axes, and somehow combine the information later, or we need to use a two-dimensional telescope array to produce two-dimensional images directly. Both techniques are discussed later in this chapter, but for now we will concentrate on one-dimensional measurements.

Examples of a fully filled and a highly diluted aperture were shown earlier in Figure 4.3. A partially filled aperture, that is, one of many segments, was also shown in Figure 4.3 (b). Recall that the diffraction pattern shown for each telescope is a cross section of the image seen by a telescope with a very small angular diameter pointed toward a star. Note that the completely filled aperture produces a simple diffraction pattern with a bright central peak and very weak sidelobes. By comparison, the highly dilute aperture produces not only a central peak but also many sidelobes that are almost as strong. If there were several stars in the field of view of a highly dilute-aperture telescope, their multiple-peaked diffraction patterns might overlap, producing a confused picture. Thus, the partially filled aperture is a compromise that gives us an easily identified bright central peak with relatively small sidelobes.

In fact, the addition of a few mirror segments between the two extreme end segments improves the image quality dramatically. By carefully choosing the placement of such segments, we can control the relative strength of the sidelobes. For example, if four mirror segments are spaced so that the center-to-center distances are 1, 2, 3, 4, 5, and 6 times the distance between the nearest pair, we essentially have six Michelson interferometers in simultaneous operation; the net result is a strong constructive interference in the central peak, accompanied by a large amount of destructive interference at points not near the peak, producing minimized sidelobes. This is a minimum-redundancy array, where the mirror-to-mirror baselines produce a maximum number of different spacings.

The Multiple Mirror Telescope (MMT) is a ground-based telescope comprising six 1.8-meter-diameter mirrors arranged at the six corners of a hexagonal cell. This pattern contains some redundancy, since opposite pairs of mirrors have the same orientation and spacing. Thus, the MMT can be considered as having a *partly dilute* aperture. Although the MMT is normally operated as a conventional telescope, it has also been successfully used as an interferometer, initially at long (submillimeter) wavelengths, and later at visible wavelengths. (For more information on the MMT, see David Latham's chapter.)

Radio astronomers have been taking advantage of the high angular resolution capabilities of separated telescopes for many years. Much of their success is because the effects of atmospheric turbulence are much weaker at radio wavelengths than at optical wavelengths, allowing phase fluctuations to be tracked and later removed from the data. The radio technique of very long baseline interferometry (VLBI) uses two or more antennas separated by thousands of kilometers, for example, between Australia and California. The operating principle is essentially the same as for the optical Michelson stellar interferometer discussed above. The Very Large Array (VLA) near Socorro, New Mexico, uses a dilute-aperture arrangement of 27 movable radio telescopes distributed along the three arms of a Y that span an area 36 kilometers across.

The COSMIC Telescope Array

The basic idea of the COSMIC telescope array is to use a modest number of medium-sized telescopes in a dilute-aperture configuration as an optical analog of the VLA. COSMIC can be configured as a two-dimensional array with either an X or a Y pattern, but it is somewhat simpler to describe a one-dimensional (I) shape. A schematic optical diagram of one possible arrangement is shown in Figure 4.5, using the same minimum-redundancy spacing as in Figure 4.3 (b). A new feature introduced here is the use of individual, afocal telescopes in place of the segments of a large single

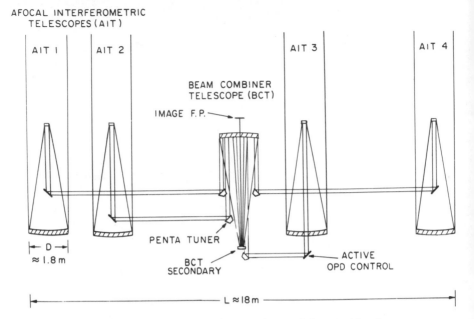

AFOCAL INTERFEROMETRIC
TELESCOPES (AIT)

AIT 1   AIT 2

BEAM COMBINER
TELESCOPE (BCT)

IMAGE F.P.

AIT 3   AIT 4

← D →
≈ 1.8 m

PENTA TUNER

BCT
SECONDARY

ACTIVE
OPD CONTROL

L ≈ 18 m

Figure 4.5: *Optical schematic diagram for a minimum-redundancy array
of approximately 18 meters' length with four collecting telescopes, each 1.8
meters in diameter. Each of the four beams is reduced in diameter and
relayed to a central beam-combining telescope, which then forms an image
in its own focal plane. The time delay of each beam is carefully adjusted
so that all light rays which started out at the same time from any selected
point on the light source (a star, say) will also arrive at the same time in
the focal plane, where an image of that point on the star will be formed.*
(Smithsonian Astrophysical Observatory illustration by Joseph Singarella)

mirror. An afocal telescope does not have a focal point, except at
infinity; instead, its function is to convert an incoming parallel
beam of light to an outgoing parallel beam of reduced diameter
at the same time it increases the angular magnification in the inverse ratio. These reduced-diameter beams of light are fed to a
central beam-combining telescope. The beam-combiner "sees" the
sky magnified from its true scale, and multiplies this further with
magnification of its own.

It is very beneficial to have more collecting elements than given by a minimum-redundancy pattern, because each additional element further improves the diffraction pattern. Because the individual telescopes are compact and optically identical, they can be manufactured efficiently. Since the cost of each additional module might be expected to decrease in an appropriately designed telescope "assembly line," it is not unreasonable to anticipate utilizing many more than just four units. If only two telescopes are inserted into the empty spaces in Figure 4.5 (and the individual mirrors are made slightly larger in diameter), we will have essentially a filled aperture, albeit still mostly in one dimension.

Figure 4.6 shows a plan view and a side view of a 36-meter COSMIC made up of two segments (each is 18 meters long and 4 meters in diameter) brought into orbit by the Space Shuttle and bolted end-to-end. The linear array has 14 individual afocal telescopes, each 1.8 meters in diameter. As the COSMIC array rotates, it sweeps out the area of a single 36-meter-diameter telescope. The large circle indicates the equivalent aperture, in terms of angular resolution, compared with that of the 2.4-meter Hubble Space Telescope, to be launched in 1986. (The technique for making a picture with COSMIC is discussed later in this chapter.)

An artist's conception of the assembly of a one-dimensional COSMIC is shown in Figure 4.7. The entire optical system can readily be designed to be bolted together in this fashion. It should also be possible to make a two-dimensional X-shaped COSMIC, as shown in Figure 4.8. Alternatively, with the more extensive assembly capabilities that will be available when the Space Station becomes operational in the mid-1990s, it may be feasible to put together a more fully two-dimensional version of the minimum-redundancy array, where telescopes are situated at seemingly random points within a large circle. This, perhaps the most powerful version of the COSMIC family, has not yet been studied in as much detail as the one-dimensional array. Thus, I will describe the one-dimensional COSMIC model and the intriguing question of how two-dimensional pictures can be obtained from a one-dimensional diffraction pattern.

Image Reconstruction

An idealized view of a central segment (slot) of a telescope lens and the corresponding elongated diffraction pattern is shown in

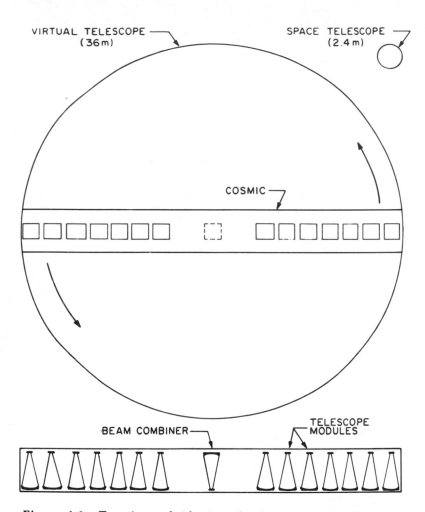

Figure 4.6: *Top view and side view of a 36-meter COSMIC telescope array with 14 collecting telescopes, each 1.8 meters square. The equivalent 36-meter circular telescope is indicated. For scale, the Space Telescope's 2.4-meter-diameter mirror is also shown.* (Smithsonian Astrophysical Observatory illustration by Joseph Singarella)

Figure 4.9. The image formed in the (x, y) focal plane is elongated perpendicular to the direction of the slot. This means that any two stars separated in the x direction by an angle corresponding to the narrow width of the elongated image will be just resolved. However, if two other stars are separated by the same angle in the y

direction, their diffraction patterns will lie nearly on top of each other, and they will be hopelessly blurred together. To separate these last two stars, the slot telescope is rotated about its *z* axis by 90 degrees and another picture is taken. In short, we can resolve

Figure 4.7: *An artist's conception of a one-dimensional COSMIC telescope array, showing the attachment of a new module by astronauts. The cutaway section shows several internal collecting mirrors, and the inset shows a typical light beam being relayed to a central beam-combining telescope. A version of the Space Station is shown in the background.* (NASA illustration)

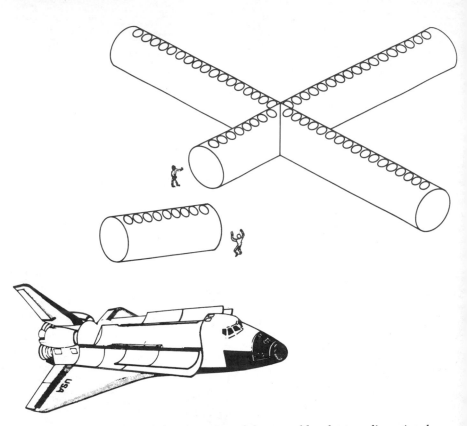

Figure 4.8: *An artist's conception of the assembly of a two-dimensional COSMIC, in a 36-by-36-meter cross. In general, two-dimensional telescope arrays, such as this cross, will allow astronomers to take pictures of fainter objects, in less time, and with less mathematical reconstruction effort, than with linear telescope arrays.* (Smithsonian Astrophysical Observatory illustration by Joseph Singarella)

any two stars separated by the minimum amount in any direction, simply by rotating the telescope about its line of sight. But what we really want is just *one* picture of an object, not a collection at different angles; this problem is solved in the following way.

When my colleagues and I developed the idea of a linear array of telescopes, we did not at have a way of combining the information that could be obtained from views taken at different rotation angles. We realized at an early stage that the mathematical technique used in the medical diagnostic technique of computer-aided tomography (CAT) could, in some sense, be applicable, since in both cases the goal is to recover details in an image that has been successively blurred or smeared strongly over a complete range of viewing angles. Eventually, we discovered a very general mathematical technique that can be applied to any aperture shape

Figure 4.9: Schematic diagram of diffraction by a slot telescope, which is similar to the case of a filled-array COSMIC. The length-to-width ratio of the aperture is the same as that of the diffraction pattern, but with the relative orientations exchanged. (Smithsonian Astrophysical Observatory illustration by Joseph Singarella)

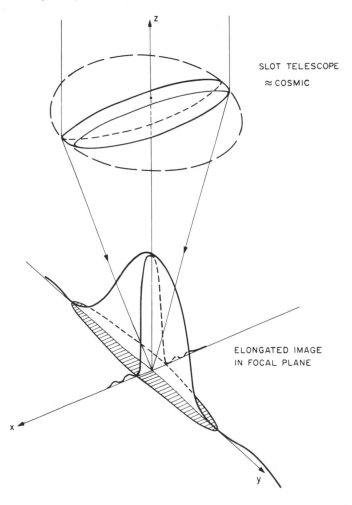

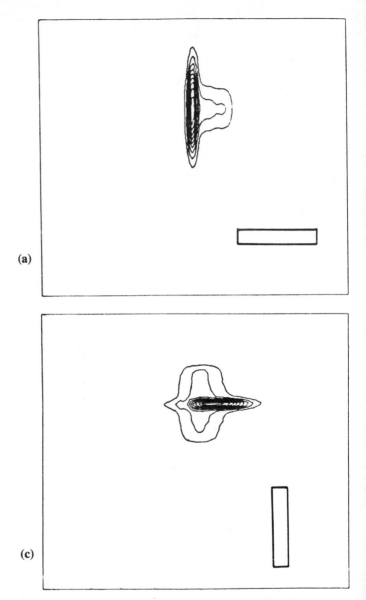

Figure 4.10: *A laboratory demonstration of the imaging of a T-shaped object through a slot aperture, such as COSMIC's filled array. The blurring effect of aperture orientations at (a) 0, (b) 40, and (c) 90 degrees is shown, along with the corresponding equivalent-diameter circular aperture to show (d) the reconstructed image.* (Harvard-Smithsonian Center for Astrophysics image processing by J. Lavagnino)

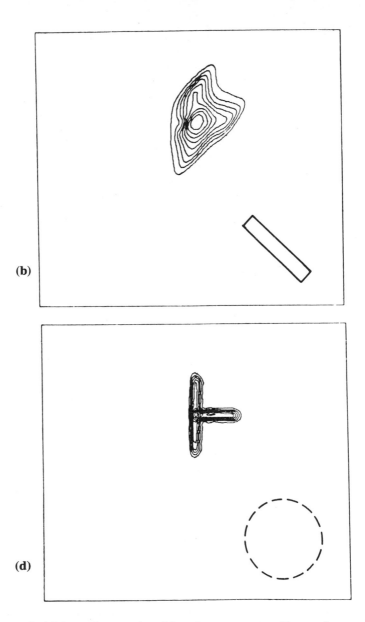

(b)

(d)

and any type of object, as long as the object does not extend beyond the telescope's field of view.

To test the image-reconstruction technique, we set up a laboratory experiment to mimic, on a small scale, the basic action of a large-scale COSMIC experiment. The light source was T-shaped, and the aperture was in a slot-telescope configuration. In Figure 4.10 (a), we see the light source as imaged through the diffracting slot oriented at zero degrees, thereby producing strong blurring in the narrow direction of the slot, as expected. Images at 40 and 90 degrees are also shown. In all, 18 images were recorded—from 0 to 170 degrees in 10-degree steps. The reconstructed image is shown in Figure 4.10 (d). The reconstruction is not simply the sum of all the individual images; rather it is a mathematically processed image that has essentially extracted the high-resolution content of each contributing image. (As a practical matter, it is important to note that the image-reconstruction technique also works quite well when there is noise in the image, or even when we have imperfect knowledge of the wavelength band or aperture geometry. This means that it is a useful "real-world" tool.)

Image Formation

There is a surprisingly strong element of commonality in the imaging techniques used in three otherwise disparate fields: radio interferometry with very long baselines (VLBI), computer-aided tomography (CAT), and the COSMIC telescope array. In each of these fields, the measuring apparatus records pictures of the object that are a tangled mixture of both sharp and blurry images. By themselves, these individual smeared images are virtually useless. However, by mathematically processing the images so as to "save" only the sharply focused parts and "ignore" the blurry parts, we eventually build up a composite picture consisting only of sharply defined elements. A brief look at each of these techniques shows this underlying commonality. (The following discussion is offered with the hope that a mathematically innocent reader will be able to understand the gist of the argument, while leaving the more technical aspects for those who are so inclined.)

VLBI: Very long baseline interferometry uses two widely separated radio telescopes that independently measure the radio emission from, for example, a quasar. The situation is sketched in Figure 4.11, where the radio beams and the analog optical beams are both shown. Each antenna receives a composite radio wave, which is simply the sum of waves from every independently emitting point on the surface of the quasar. This wave is recorded at

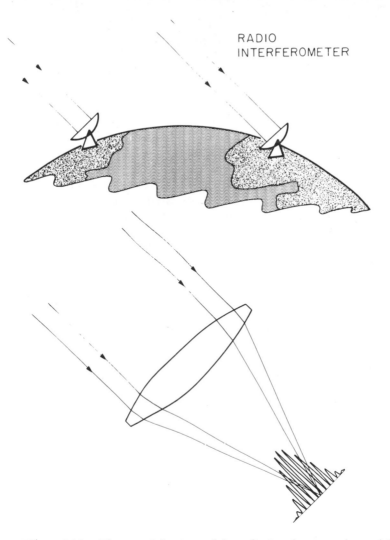

Figure 4.11: *The essential nature of the radio interferometer (upper) is to combine two beams with an adjustable time delay so that the fringe visibility can be measured. For comparison, an optical Michelson stellar interferometer (lower) produces a fringe pattern from which the visibility is determined from the peak-to-valley variation.* (Smithsonian Astrophysical Observatory illustration by Joseph Singarella)

each antenna, along with timing information. Later the recorded signals are combined in a computer. If the waves from a point on the quasar are combined with zero relative delay, then constructive interference will occur and the intensity from that point will be a maximum. If the same waves are combined with exactly one-half wavelength of delay, then destructive interference will occur and the intensity from that point will be zero, which is a minimum. The relative change in the combined signal is the *visibility*, in this case, 100 percent. In the optical analog case, the visibility is the relative change in going from the peak to the valley of the interference fringe pattern shown in Figure 4.11, again 100 percent.

In both the optical and radio cases, if there is a second nearby source in the sky, it will produce its own interference fringes, generally with different peak and valley locations, thereby reducing the visibility of the total fringe pattern to less than 100 percent.

The *phase* of the measured visibility is determined by noting how much relative delay is needed to produce a maximum intensity of interference. For example, if the maximum occurs at zero relative delay, then the phase is zero degrees, but if it occurs at, say, one-quarter wavelength of spatial delay, then the phase is 90 degrees.

It is helpful to visualize the radio fringe pattern as an optical analog by drawing it projected onto the sky. In Figure 4.12 (upper left-hand corner) a radio object consisting of a blob and three pointlike sources is sketched, using an (x, y) coordinate system. The fringe pattern from one of these pointlike sources has been superposed as a bar pattern. The orientation and spacing of the bars is determined by the baseline orientation and spacing of the radio telescopes. Every other point in the source will contribute its own fringe pattern to the total, but offset from the indicated fringe pattern by the same amount as the source is offset.

The visibility measurement is conveniently represented by a number in (u, v) space, along with its mirror image at $(-u, -v)$. Here u is the spatial frequency of the bar pattern in the x direction (i.e., u = number of bars per degrees across the sky in the x direction), and v is the spatial frequency in the y direction. If we

Figure 4.12: (Clockwise from upper left) In radio interferometry, sweeping the fringe pattern across the source gives one point (and its complex conjugate) in the Fourier plane. Summing, inverse transformation, and the use of the computer program CLEAN give the reconstructed image. (Smithsonian Astrophysical Observatory illustration by Joseph Singarella)

RADIO INTERFEROMETER

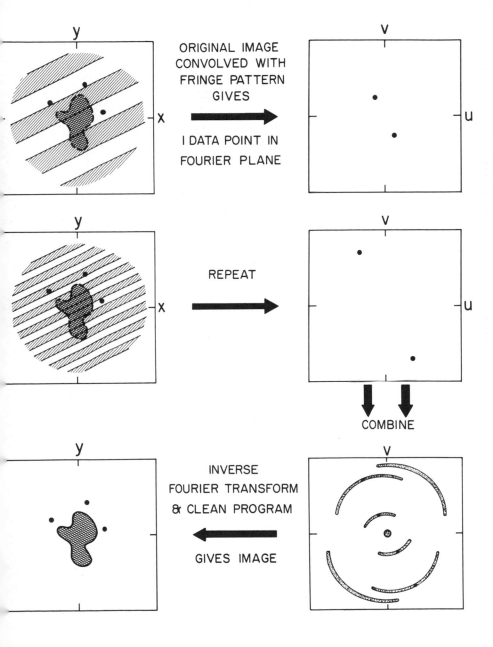

ORIGINAL IMAGE
CONVOLVED WITH
FRINGE PATTERN
GIVES

I DATA POINT IN
FOURIER PLANE

REPEAT

COMBINE

INVERSE
FOURIER TRANSFORM
& CLEAN PROGRAM

GIVES IMAGE

CAT SCANNER

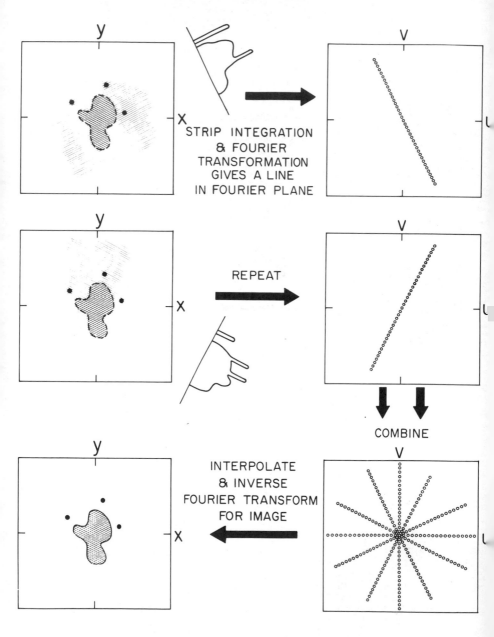

STRIP INTEGRATION
& FOURIER
TRANSFORMATION
GIVES A LINE
IN FOURIER PLANE

REPEAT

COMBINE

INTERPOLATE
& INVERSE
FOURIER TRANSFORM
FOR IMAGE

include the phase information, then the visibility is represented by a complex number instead of just a real number.

A remarkable theorem—from the French mathematician J. Fourier—says that any two-dimensional picture can be duplicated by adding up a number of adjustable-intensity sinusoidal wave patterns that have both a variety of wavelengths and a variety of directions. Any one of these waves can be visualized as a bar pattern, with fuzzy edges on the bars, set at an angle to the (x, y) axes. To reproduce the picture requires both very wide bars and very narrow bars, all angled at many different orientations. To keep track of the prescription for the collection of bar patterns needed, we can use a new coordinate system, the (u, v) plane, where wide bars are represented by points near the origin ($u = 0$, $v = 0$), and narrow bars by points far from the origin. The direction of the point in the (u, v) plane is the same as the direction of the perpendicular line between bars in the (x, y) plane. To complete the description, the intensity or strength of each bar pattern is represented by a number (or height in a three-dimensional plot) at the appropriate point in the (u, v) plane. The collection of all these points is called the *Fourier transform* of the image, and is a complete catalog of all spatial frequency waves needed to reproduce the original picture. The reverse process is called the *inverse Fourier transform*, which gives us back the original picture.

If the object is ascending in the sky, then, at a later time, the VLBI baseline is increased so the fringes move closer together, and we are able to determine two more points in the (u, v) plane, and so on. After all the data are accumulated, we have a partially filled (u, v) plane, shown in the lower right-hand box of Figure 4.12. The "tracks" in this plane are determined by the position of the object in the sky, the location of the telescope, and the rotation of the Earth. The inverse Fourier transform could yield a picture of the object if the (u, v) plane were filled, but, since it is not completely filled, we get instead an image (not shown) with many

Figure 4.13: *(Clockwise from upper left) In a computer-assisted tomography (CAT) scan, a strip integration and Fourier transformation give points in the spatial frequency plane. Summing, interpolation, and inverse transformation give the reconstructed image.* (Smithsonian Astrophysical Observatory illustration by Joseph Singarella)

false features. These unwanted features can be removed with the widely used computer program CLEAN, which subtracts a series of diffraction patterns such as would be produced by an isolated single point source in the sky, and thus generates an estimated brightness pattern that represents the "real" sky. This recovered image is shown in the lower left-hand panel.

CAT: In diagnostic medicine, the X-ray CAT scanner sends a large number of narrow pencil beams of X rays through a body section to be measured by an equal number of X-ray detectors on the other side. The loss of intensity is noted for each pencil beam, and corresponds to the total density of absorbing matter along the ray. This is clearly analogous to integrating the brightness of an astronomical source along strips in the sky, as shown in Figure 4.13. In fact, the early development of the mathematical technique shown here was for astronomy, not medicine. Calculation of the Fourier transform of this strip integral produces simultaneously all the possible sine and cosine components that had to be built up step-by-step in VLBI. These complex-valued points, representing amplitude and phase, are stored in the (u, v) plane.

In the CAT diagnosis, this process is repeated at different view angles. The (u, v) plane is successively filled in with measured points. These values are then interpolated to a convenient grid, and the inverse Fourier transform taken, which directly yields the reconstructed object. (In practice, the CAT image is reconstructed by a mathematical procedure that is different from that presented here, although the two are totally equivalent.)

COSMIC: As shown in Figure 4.14, the image recorded by the COSMIC telescope array is closer to being a true picture than is the case for either VLBI or CAT, since it is much less severely smeared by the observing apparatus. Thus, the Fourier transform of this image immediately fills in a substantial fraction of the (u, v) plane. By rotating the aperture to different orientations, the needed information fills in the remainder of the (u, v) plane.

Figure 4.14:   *(Clockwise from upper left) In COSMIC image reconstruction, the original image is partially blurred in one direction only, so the spatial frequency plane is partly filled. Weighted summing, filtering, and inverse transformation give the reconstructed image.* (Smithsonian Astrophysical Observatory illustration by Joseph Singarella)

COSMIC OR SLOT TELESCOPE IMAGING

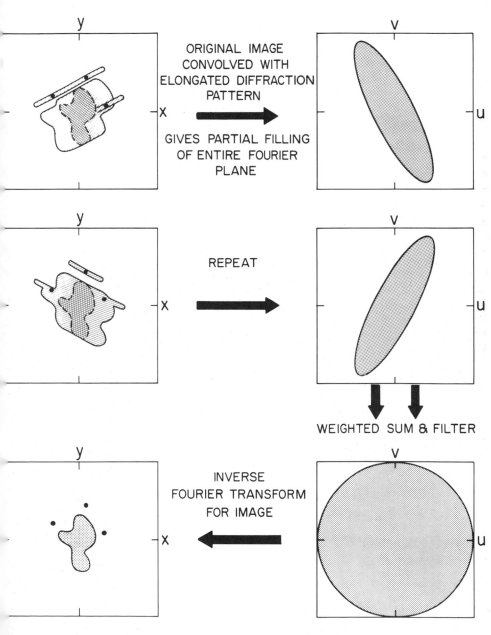

ORIGINAL IMAGE
CONVOLVED WITH
ELONGATED DIFFRACTION
PATTERN

GIVES PARTIAL FILLING
OF ENTIRE FOURIER
PLANE

REPEAT

WEIGHTED SUM & FILTER

INVERSE
FOURIER TRANSFORM
FOR IMAGE

Just as the one-dimensional array fills in a strip in the (u, v) plane, a two-dimensional array can provide information over the entire (u, v) plane in one exposure. We see again, from this different point of view, that a two-dimensional array will not have to be rotated in order to provide complete (u, v) coverage. From this point of view, too, a minimum-redundancy array (one- or two-dimensional) provides the most uniform coverage of the (u, v) plane, when compared to any other type of array having the same overall diameter and collecting area.

As a one-dimensional array is rotated, the new (u, v) plane information sometimes overlaps previous information, and sometimes does not. This problem is addressed by first multiplying each contribution by a special weighting function, which is simply the response to a single point source in the sky. After adding all contributions, the weighted sum is "unweighted" by the sum of the above weights, then "reweighted" by the shape function corresponding to a large circular aperture with diameter equal to the length of the array, and, finally, an inverse Fourier transform is taken to directly yield an image.

Note that in both VLBI and CAT the spatial frequency information was extracted and saved in the (u, v) plane. Only when all the available information was accumulated was it possible to invert the process, and thereby reconstruct an image corresponding to the collection of spatial frequencies. Although the COSMIC image-reconstruction technique follows this same process, it is a more general mathematical procedure that includes as special cases both VLBI and CAT.

The COSMIC Approach

There appear to be no major technical roadblocks to building a COSMIC telescope array. Although some sophisticated design and fabrication techniques will be required, all are either currently available or soon to be developed. For example, the mechanical structure should be of a material with a low coefficient of thermal expansion, like the graphite epoxy trusses built for the Hubble Space Telescope. The mirrors should be very lightweight to minimize the mass to be lifted by the Space Shuttle; fortunately, the development of lightweight mirrors is currently an area of very active study. Computer-aided mirror-polishing methods, which promise to reduce the cost and time of manufacture of large mirrors, are being developed. The optical system needs to be actively aligned on a periodic basis to compensate for thermal and gravitational flexure; here again, active mirror technology is currently under rapid development in many laboratories. Two-dimensional photon-counting detectors are needed for faint sources, and these, too, are proliferating in many laboratories. Moreover, rather standard spacecraft components are becoming available that will cut the time and cost associated with basic utility functions, such as power, telemetry, and aspect control.

The original linear array configuration of COSMIC was a direct outgrowth of the shape of the Space Shuttle bay. With the vastly increased assembly capabilities expected with the development of a permanent orbiting platform, or Space Station, we have before us the tantalizing prospect of building a sparsely filled two-dimensional array. The primary advantage of such a telescope is that it would need little or no rotation (depending upon the array configuration) in order to build up an image. To fulfill this potentially very important function, the Space Station must have facilities for the assembly, deployment, and checkout of large astronomical instruments.

To illustrate the tremendous scientific potential of COSMIC, consider a 36-meter linear array of telescopes, containing 14 individual collecting telescopes, each 1.8 meters square. Compared to the Hubble Space Telescope, COSMIC will have 10 times more collecting area and 15 times more angular resolution. Figure 4.15 shows a graphic example of how this gain in angular resolution will allow us to use COSMIC to image individual stars in globular clusters in many neighboring galaxies. With this kind of resolution, COSMIC should be able to address many of the major questions of modern astrophysics, including some that will certainly remain unanswered long after the launch of the Hubble Space Telescope.

Will the universe continue to expand forever? COSMIC can determine cosmic deceleration by applying the angular-size/redshift test to the cores of the brightest galaxies in clusters—objects now assumed to be "standard measuring rods." This test works as follows: suppose that all galaxies of a certain type have identical diameters. In Euclid's universe, the more distant galaxies appear smaller in angular size: objects twice as far away appear to be half-

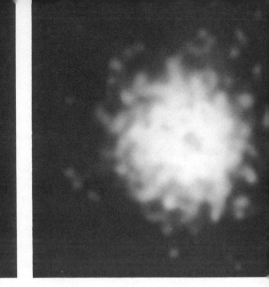

(a) (b)

Figure 4.15: *To illustrate the angular resolution capability of COSMIC, three images are shown here of a globular cluster (M3) in our galaxy as it would appear if in another galaxy 1,000 times more distant. Panel (a) shows the globular cluster as it would be seen by the Hubble Space Telescope, that is, totally unresolved. Panels (b) and (c) show the same object as it would be seen by an 18-meter-long COSMIC and a 36-meter-long COSMIC, respectively, showing individual stars clearly resolved.* (Harvard-Smithsonian Center for Astrophysics image processing by J. Lavagnino)

size. The same rule holds in Einstein's universe for nearby distances, but at very large distances, where galaxies are receding at a substantial fraction of the speed of light, the rule breaks down. Instead, at a certain distance, the most distant galaxies start to appear as nearly constant in size—or even larger! This breakdown in the Euclidean rule is a very powerful first-order test of cosmological models.

What is the value of the Hubble constant? COSMIC can extend the extragalactic distance scale by observing Cepheid variable stars in relatively nearby clusters of galaxies, such as the Virgo and Pegasus clusters, and by using globular cluster systems in more distant galaxies, including those as far away as the Coma and possibly Hercules clusters. Such observations are important for studies of large-scale dynamics of clusters of galaxies as well as the determination of the Hubble constant.

(c)

What is the "fuzz" around quasars? COSMIC can investigate the nature of the diffuse emission seen around certain quasars to see whether it represents an underlying galaxy, and if so, what type. This is one step in the direction of understanding the physics of quasars.

What lies at the centers of active galaxies? COSMIC can probe the structure of the regions surrounding the nuclei of Seyfert galaxies down to sizes the equivalent of about a light-year, that is, the scale on which broad-line spectral variations are seen.

Is there evidence for black holes? COSMIC can study the nuclei of ordinary galaxies with suspected massive black-hole centers to see if the gravitational potential is truly pointlike. In addition, COSMIC can examine the so-far unresolvable central regions of globular clusters for evidence of mass distributions indicative of either black holes or self-gravitation effects.

Can we see features on the surfaces of stars? COSMIC should be able to image the surfaces of nearby supergiant stars, where complex motions involving expansion, contraction, and shocks are suggested by the maser emissions seen with VLBI. These surface features might include: brightness inhomogeneities due to large convective cells, overall shape changes due to photospheric motions, chromospheric spectral emission in strong hydrogen and calcium lines, and large-scale pulsations. COSMIC can also image nearby main-sequence stars to detect their rotation axes, roughly follow latitude distribution of surface features as a function of the activity cycle, search for differential rotation, and search for large-scale magnetic activity.

How did the Solar System form and evolve? COSMIC can make a detailed survey of Pluto for radius, rotation rate, and axial orientation, and determine the orbit of Charon; observe the Saturn ring system with a resolution approaching that achieved by the best Voyager 2 images; make time-dependent studies of volcanic activity on Io and features in Jupiter's atmosphere, again with resolution comparable to that obtained by Voyager 2; resolve cometary nuclei and monitor activity of the inner coma, including the formation and location of jets and possible breakup of the nucleus; and measure hundreds of asteroids for size, shape, rotation, and surface features.

A Family of Interferometers in Space

COSMIC is only one member of a family of potential optical interferometers in space. If we include ultraviolet and infrared but not radio interferometers, then, by my count, the family currently includes at least a dozen distinct interferometers that have been proposed by scientists from a half-dozen countries. (Here we limit our discussion to interferometers that can provide imaging information, thereby excluding astrometric devices.)

The interferometer family can be divided into two groups, rigid and floating. The rigid interferometers have all their optical elements rigidly connected by a single framework: COSMIC is a rigid interferometer. The floating interferometers comprise two or more free-floating spacecraft with no stiff mechanical interconnections. A brief look at these concepts will give some idea as to the variety of potential configurations, as well as the advantages and corresponding disadvantages. The reader should be aware that these are all relatively young concepts, with ages ranging from a few years to a few months! The ideas will surely evolve in time as they are studied and refined.

Figure 4.16: An artist's conception of SAMSI as it moves in a spiral pattern, propelled by small thrusters. (Harvard-Smithsonian Center for Astrophysics illustration by Beryl Langer)

SPACECRAFT ARRAY FOR MICHELSON
SPATIAL INTERFEROMETRY
(SAMSI)

A system called Optical Aperture Synthesis in Space (OASIS) is similar to COSMIC in that it is a rigid interferometer consisting of a two-dimensional array of collecting telescopes with comparable baseline, angular resolution, collecting area, and number of telescopes. OASIS differs mainly in that it does not make direct images, but instead forms fringe patterns between all pairs of telescopes. The visibility of the fringe patterns is measured and images are computed later, using techniques borrowed from VLBI image reconstruction. Because fringes are measured at many different nearly monochromatic wavelengths, the alignment accuracy of each mirror is relaxed to a few wavelengths, compared to a fraction of a wavelength for COSMIC. However, this also means that the ultimate sensitivity of OASIS is limited to relatively bright objects, essentially the price paid for relaxed alignment tolerances.

SAMSI (Spacecraft Array for Michelson Spatial Interferometry) and TRIO (named for its original design of three components) are both free-floating interferometers, with two (or more) collecting telescopes and a central beam-combining station (Figure 4.16). Both have variable baselines up to about 10 kilometers, giving resolutions of 10 microarcseconds. Both measure fringe visibilities rather than forming direct images. As with OASIS, the alignment tolerances are relaxed to a few wavelengths, so that the attainable limiting magnitudes are fairly bright. The fields of view will also be small, on the order of 0.1 arcsecond. SAMSI is designed for Earth orbit, either with slightly inclined intersecting orbits, or with active positioning control, using chemical thrusters. TRIO is different in that it is stationed at a fixed point with respect to the Earth–Moon system—the Lagrangian L5 point. Here the tug of gravity is sufficiently weak so that the propulsive effect of photon pressure from the Sun is being considered as an alternative to active on-board thrusters; in short, the solar wind will push against a space sail to maneuver the satellite.

Conclusion

An array of optical telescopes in space, such as COSMIC, offers the great advantage of achieving both high angular resolution imaging and large collecting area in the same instrument. The linear array version of COSMIC can have its individual collecting telescopes configured in a minimum redundancy pattern. However, a filled, or nearly filled, array has the additional advantages of small sidelobes and high tolerance to the loss of any individual telescope. Fully two-dimensional images can be readily constructed from observations covering 180 degrees of rotation of a linear array about its line of sight. The presence of a large semipermanent space station orbiting the Earth would certainly simplify the deployment, assembly, and servicing of COSMIC. Moreover, an even more powerful instrument—a two-dimensional COSMIC—might be assembled at such an orbiting platform. Much of the required technology to build COSMIC is currently available, and the remaining technology should be developed in the very near future. A 36-meter array having an angular resolution of 3 milliarcseconds in the visual band and a collecting area about 10 times that of the Hubble Space Telescope would allow scientific investigations that are now well beyond the capabilities of any available or planned telescopes, particularly those requiring imaging of faint, complex objects.

Antenna Earth

Astronomy with an Intercontinental Array of Radio Telescopes

MARK J. REID

Decades ago, astronomers recognized that many interesting types of astronomical objects had angular sizes much smaller than the 1-arcsecond resolution typical for optical telescopes. Restricted to visible light, however, optical astronomers could only imagine the wealth of fascinating astrophysical phenomena to be explored with higher angular resolution telescopes. Rapid technological innovation after World War II, including the ability to carry instruments above the Earth's atmosphere, opened new windows in the electromagnetic spectrum, from the very short-wavelength X rays to the long-wavelength radio waves, but the problem of improving angular resolution remained considerable.

Indeed, for the long-wavelength end of the spectrum, the situation seemed particularly bleak. Because the size of the smallest object that can be resolved with a telescope generally is proportional to the wavelength of the radiation observed, radio astronomers were at a great disadvantage. Radio waves are about 10,000 times longer than visible light. Hence, to achieve the same resolving power, a radio telescope must be about 10,000 times larger than its optical counterpart. To build such a telescope—about 10 kilometers in diameter—clearly was, and still is, far beyond our technological capabilities.

However, in the late 1950s, scientists in England found that the signals received by several small radio antennas could be transmitted over cables and then "coherently" combined to make images with resolution comparable to optical photographs. This technique, called radio interferometry, is largely due to Martin Ryle, who later received the first Nobel prize for a contribution in astronomy. By spreading a handful of small antennas first over a one-mile area,* and later over 5 kilometers, radio astronomers were able to link them together and "synthesize" a telescope with an aperture equivalent to the distance between the most widely separated members of the array and, more important, with angular resolution approaching 1 arcsecond (see preceding chapter). However, by the early 1960s, it looked as though practical limitations on the length of the cables used to transmit radio signals to the central interferometer building would restrict the size of a synthesized telescope to about 10 kilometers. At that time, it therefore seemed unlikely that any ground-based astronomers would crack the "1-arcsecond barrier."

In response to this challenge, radio astronomers in Canada and the United States surprised the astronomical world by extending the distance between antenna components of a radio interferometer, not just to a few tens or even a few hundreds of kilometers, but to thousands of kilometers. This was accomplished in the late 1960s by replacing the cables linking the interferometer components with magnetic tape recorders and coordinated precise time standards. The recorded signals from a celestial source received separately by each antenna were later brought to a central computer where, with the help of the basic time reference, they were combined to produce a pattern of interference fringes that could be reconstructed as a single image. The new technique, dubbed very long baseline interferometry (VLBI), increased the angular resolution available to astronomers nearly a thousandfold. Currently, radio astronomers are producing images of radio sources with an angular resolution of 0.0003 arcsecond. Were the human eye to have this power, one could read these words from a distance of about 3,000 miles!

---

* This first attempt is generally called the "one-mile antenna."

Such unprecedented angular resolution is currently being used to study objects such as the astounding energy sources in the centers of quasars and radio galaxies, the enigmatic sources of natural maser (i.e., radio-frequency laser) radiation from both newly forming and dying stars, and many other exotic astronomical phenomena. However, the immediate future holds the promise of a 10-element intercontinental array of antennas dedicated solely to VLBI observations; and, following the advent of this very long baseline array later this decade, the further extension of radio interferometry with antennas in space may produce angular resolution better than 0.0001 arcsecond.

Quasars and Radio Galaxies

If one travels to a good dark-sky site, such as a typical optical observatory on a high mountain away from city lights, the evening sky reveals a bright band of light we call the Milky Way, a portion of the vast collection of some 100 billion stars that make up our galaxy. More stunning, if one takes a "deep" photograph of the sky with a large optical telescope, for each of the 100 billion stars in the Milky Way, one is likely to see another galaxy, itself containing about 100 billion stars!

Galaxies also cluster together, as shown in Figure 5.1, forming the greatest conglomerations of materials in the universe. Typically, at the center of a cluster of galaxies, there is a particularly large elliptical galaxy that emits a fantastic amount of energy. However, when looked at in radio "light," instead of visible light, these central galaxies often appear quite different in form. With visible light, one sees only a bright group of stars; with radio light, the picture is dominated by two giant lobes hundreds of times the size of the galaxy and symmetrically placed about it. Figure 5.2 shows an example of a radio galaxy with its lobes as imaged with a radio interferometer.

Giant radio lobes have a staggering amount of energy tied up in fast-moving particles and in magnetic fields. The minimum amount of energy contained in these structures is equivalent to the conversion of more than one million Suns *entirely* into energy! (Using Einstein's famous formula, this can be expressed as: $E = mc^2 \times 1,000,000$, where m is the mass of the Sun and c is the speed of

Figure 5.1: *A cluster of galaxies toward the constellation Virgo. Each galaxy contains about as many stars as the Milky Way (e.g., 100 billion).* (Kitt Peak National Observatory photograph)

light.) By contrast, a hydrogen bomb converts only about a pound of material to energy, an infinitesimal amount of mass compared to the Earth, which itself is only about a millionth the mass of the Sun.

Recently, images of radio galaxies produced by the 35-kilometer,

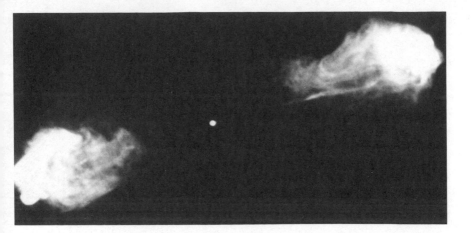

Figure 5.2: *An image of the radio-frequency emission from a distant galaxy toward the constellation Cygnus, showing the double-lobed pattern typical of the radio galaxies. The galaxy, which supplies the energy for the radio lobes, is the small bright spot in the center of the picture.* (Photograph from Peter Scheuer, Robert Laing, and Richard Perley; VLA, National Radio Astronomy Observatory)

Y-shaped array of radio telescopes called the Very Large Array (Figure 5.3) showed narrow filaments of emission connecting the giant radio lobes to the center of the galaxy. These filaments appear to be channels through which energy is transported to the radio lobes from the nucleus of the galaxy—its central "engine." Optical telescopes with arcsecond resolution see the nucleus of the galaxy just as a point of light. Thus, only by using very long baseline interferometry can we hope to resolve—and understand—the central mechanisms that power radio galaxies.

In fact, VLBI observations of radio-galaxy nuclei show them to be very compact. Typically, the radio emission comes from a region less than one light-year in diameter. Astronomically speaking, this is a tiny distance—less than that from the Sun to the nearest star. The realization that such incredible amounts of energy are produced in such minuscule volumes of space is one of the pieces of information that leads astronomers to believe the cores of radio galaxies may contain gigantic black holes.

The discovery of radio galaxies and their very compact nuclei was exciting, but radio astronomy was responsible for the discovery of even more energetic and puzzling objects—the quasi-stellar radio sources, or *quasars*. As the name suggests, a quasar appears as a point of light, resembling a star when viewed optically, but, in radio (and other wavelengths), emitting as much energy as an entire galaxy. Moreover, quasars appear to be moving away from us at speeds that are a significant fraction of the speed of light. According to Hubble's Law, the velocities of objects receding from us increase with their distance; thus, quasars appear to be the most

Figure 5.3: *The very large array (VLA) radio telescope of the National Radio Astronomy Observatory in New Mexico. The instrument consists of twenty-seven 25-meter-diameter antennas. They are placed on a Y-shaped rail system and can be moved so that they span a 35-kilometer-diameter area. This instrument achieves better than 1-arcsecond resolution at centimeter wavelengths.* (National Radio Astronomy Observatory photography)

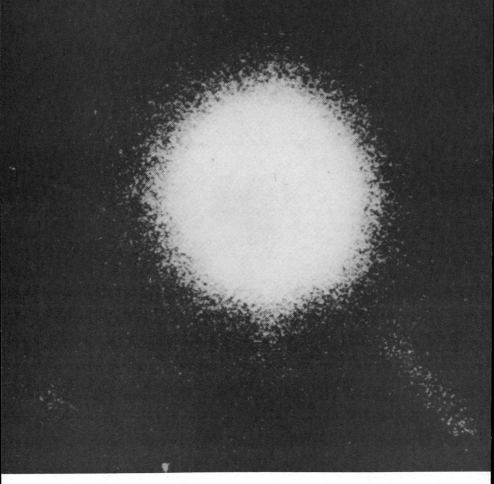

Figure 5.4: *An enlarged photograph of the quasar 3C273. The source displays a bright pointlike core (overexposed), and a dim jetlike wisp extending 20 arcseconds toward the southwest (lower right).* (Harvard College Observatory photograph)

distant objects in the universe. And, to be seen at these great distances, they must also be the most luminous sources in the universe.

One particularly intriguing quasar is called 3C273. (The designation stands for the 273rd source discovered in the third Cambridge University survey of the sky in radio waves.) A greatly enlarged optical photograph of this source (Figure 5.4) shows 3C273 is composed of a bright compact spot (overexposed on this print) and a faint wisp, or jetlike structure, extending for about 20 arc-seconds toward the lower right of the figure.

When viewed with an angular resolution of about 0.001 arc-second, the compact component of 3C273 displays remarkable structure. Indeed, VLBI maps of the milliarcsecond structure of this compact component show it to be composed of a pointlike bright spot that blends into a jetlike extension. The orientation of this jet mimics that seen in the optical photograph, except on a scale ten thousand times smaller. Thus, as is the case for the radio galaxies, all the energy emitted by the quasar can be traced back along channels to an exceedingly small central region. But here the total energy output is even more extraordinary: 100 billion times that of the Sun, and yet all apparently emitted from a volume less than from the Sun to the nearest star.

New developments in astronomical instrumentation and observing techniques have almost always been rewarded with the discovery of some new facet of the physical universe. The advent of very long baseline interferometry proved no exception. While observing some particularly bright quasars (3C273 included) with a VLBI array several times over a few months, astronomers noticed that the data consistently supported a suggestion that the quasars were expanding—expelling huge blobs of emission—at a rate much greater than the speed of light!

Much controversy surrounded these startling results. Was Einstein's theory of relativity wrong? Could objects really travel faster than the speed of light? Or was the Hubble Law, linking distance to recessional velocity, grossly in error so as to make quasars nearby objects? Most physicists were reluctant to abandon the theory of relativity, since it had passed numerous detailed experimental and observational tests. Similarly, most astronomers were not likely to modify substantially the Hubble Law so as to "shrink the universe" by a factor of 10. (Such a reduction in size would correspondingly make the age of the universe "uncomfortably" shorter than geologic evidence indicates for that of the Earth!)

When the so-called superluminal motions in quasars were discovered, VLBI was still in its infancy. It was not possible to produce actual images as is done today; thus, the observational data were criticized for not *uniquely* determining the structure of the emission source. However, as techniques improved, evidence for apparent superluminal motion increased, until images of quasar 3C273 taken

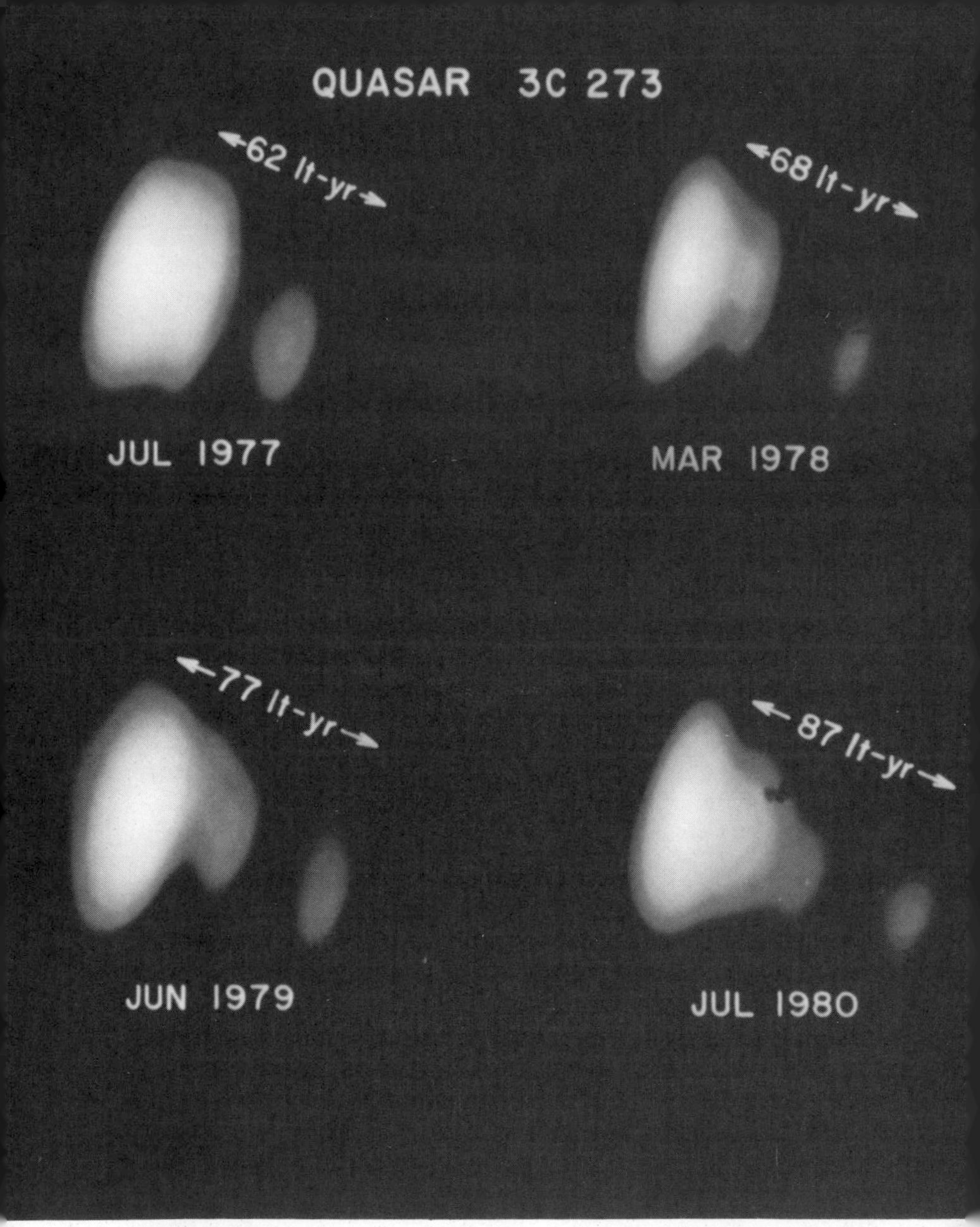

Figure 5.5: *Images of the central component of the quasar 3C273 at radio frequencies. These images have milliarcsecond resolution and were made with a VLBI array of radio telescopes spanning two continents. The apparent expansion rate of the source over the four time periods observed greatly exceeds the speed of light!* (Image from the Owens Valley Radio Observatory, California Institute of Technology)

over several years (Figure 5.5) clearly showed the expansion.

While there is certainly not unanimity among astronomers as to the explanation for the observations, one theory seems most plausible. Einstein's theory contains some subtle characteristics implying that, under certain conditions, objects can *appear* to move faster than the speed of light. This "relativistic illusion" occurs only when sources of light are moving almost directly toward us at nearly (but still slower than) the speed of light. Thus, scientists speculate they may be seeing great bursts of emission shot from the cores of quasars into our line of sight. What kind of incredible powerhouse inside the quasar could produce this energetic outpouring still remains a mystery, however.

Masers

An image of our own Milky Way taken by an observer in another galaxy would probably reveal a spiral structure dotted with many bright knots of emission (much like the galaxy shown in Figure 5.6). These knots are regions of recent star formation. Specifically, these are sites where stars tens of times as massive as the Sun have recently formed. They are now emitting copious amounts of en-

Figure 5.6: *A spiral galaxy thought to be similar to the Milky Way.* (Smithsonian Astrophysical Observatory photograph by Rudolph Schild)

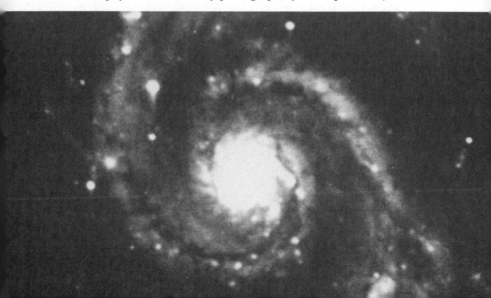

ergetic (ultraviolet) light, and are ionizing the gases of their placental environment. Near the brightest and most compact of these ionized regions (called HII regions) one commonly finds intense emission from molecules such as water (H_2O) and hydroxyl (OH). When radio astronomers first detected this molecular emission they were quite puzzled by its characteristics—high polarization, unusual spectral line strengths, and very narrow spectral line widths. The first VLBI observations of these sources demonstrated that, were the radiation to come from "starlike" material, the material would have to be at a temperature of about 100 billion degrees. This was far too hot for normal emission processes and suggested that some coherent process of radiation amplification—a natural maser—must be involved.

Recent VLBI observations of OH masers have been very successful in mapping the emission. The best-studied hydroxyl source is associated with an ultracompact HII region called W3 (OH). Figure 5.7 shows the locations of some 70 OH maser spots relative to the ionized emission from free electrons in the clouds of the compact HII region. This map shows that the OH masers congregate in clusters that are about 100 AU in diameter (an AU, or astronomical unit, is the distance from the Sun to the Earth). Each cluster contains a mass roughly equal to that of the planet Jupiter. What one "sees" when observing interstellar masers, then, are the ends of cosmic amplifiers. Each acts like a giant transmitter, roughly the size of our Solar System, to broadcast intensified signals from a newly formed star.

Figure 5.8 shows two of these OH maser clusters at a magnification factor of about 100, obtained with VLBI's "3,000-kilometer lens" and displaying the OH emission as a photograph of radio intensity. The Doppler, or redshift, velocities of individual components are indicated by the numbers (in units of kilometers per second with respect to the average velocity of material near the Sun). Since spectral observations have frequency as well as spatial information, one can display the four-dimensional information (intensity, frequency, and x, y positions on the plane of the sky) using color to represent radio frequency in a manner similar to the eye's perception of visible light. When this is done, one creates a radio photograph equivalent to what the human eye would see were it (1) sensitive to radio waves, (2) the size of the United States, and (3) capable of spectral resolution of one part in a million.

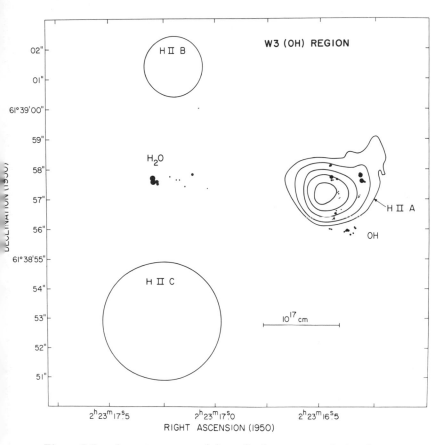

Figure 5.7: *A contour map of the radio-frequency emission from an optically obscure region of star formation. The black "spots" indicate the locations of hydroxyl (OH) and water vapor (H_2O) maser sources.* (Smithsonian Astrophysical Observatory illustration by Joseph Singarella)

One of the fields presented in Figure 5.8 has two maser components that fall precisely on top of each other, but that have velocities of 3.4 and 6.4 kilometers per second. These features, which are oppositely polarized, result from splitting of the hydroxyl line by the presence of a magnetic field, a phenomenon known as the *Zeeman effect*. In W3 (OH), nearly half of the frequency spread in the spectrum is due to Zeeman splitting rather than real motions. (The magnetic field strength determined from the Zeeman splitting

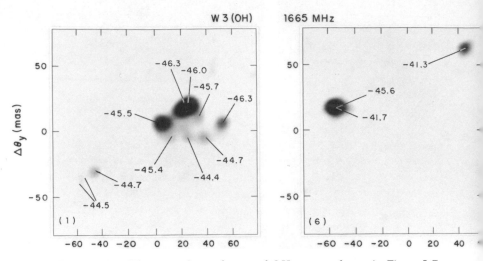

Figure 5.8: *Blowups of two clusters of OH masers shown in Figure 5.7. These images were made with data from a transcontinental VLBI telescope. The angular scales are in milliarcseconds (i.e., 0.001 arcsecond units). The numbers indicate the radial velocities of identifiable maser features in units of kilometers per second.* (Smithsonian Astrophysical Observatory illustration by Joseph Singarella)

is about 5 milligauss, or about one one-hundredth of the field strength of the Earth.)

After subtracting the apparent velocity shifts due to Zeeman splitting, one finds that almost all clusters of OH masers in a given HII region have the same center-of-mass velocity. In W3(OH), this velocity is redshifted by about 5 kilometers per second with respect to the velocity of the central star (-50 kilometers per second) as measured by the radio-frequency spectral lines produced by the recombination of electrons with protons to form hydrogen atoms in the HII region. Since the ionized material is opaque at the low frequency of the OH transition (1,665 megahertz), the OH masers cannot be seen through the HII region. Thus, they must be in front of the HII region. Moreover, the locations and velocities of the OH masers relative to the HII region indicate that the masers are falling inward toward the central star. In short, by observing the OH masers, we seem to be probing the remnant material out of which the newly formed star was created.

We are fortunate when studying maser emission that the OH molecule has other observable transitions. One important transition occurs at a frequency of about 6,000 megahertz. Although very few telescopes are outfitted with receivers capable of tuning to this frequency, some primitive VLBI observations of W3 (OH) have shown very clear patterns of Zeeman splitting across the entire spectrum. With more detailed observations of this OH maser emission (i.e., determining the circular and linear polarization characteristics), one can determine the three-dimensional magnetic field arrangement in a source. Such information is crucial for the understanding of star formation. In W3 (OH), for example, the pressure of the gas inside the HII region is roughly equal to both the pressure of the infalling neutral gas and the pressure exerted by the magnetic field on the region. In short, many forces may play important roles in the collapse of interstellar material to form stars.

The strongest molecular maser sources are water-vapor (H_2O) masers—and some are particularly spectacular. For example, a water maser in the Orion region has been observed to flare up by a factor of more than 1,000 in the (astronomically speaking) short period of a few months. And it reached a peak strength nearly a million times greater than most other sources! Intense water masers like this one may emit, in a single radio channel only 50 kilohertz wide, energy equal to a large fraction of the entire luminosity of the Sun. Translated into terrestrial terms, if such a source were a radio station, it would be broadcasting with nearly 10^{20} megawatts of power, or about the energy needed to light a trillion-trillion 100-watt electric light bulbs!

Water masers found in regions of star birth appear to be associated with earlier phases of a star's formation than are the OH masers. However, in common with the OH masers, water masers do form in groups, sometimes comprising hundreds of spots. Each spot in this cluster has a distinct velocity and position on the sky, thus allowing measurement of the relative motions of individual maser spots on the plane of the sky. For example, water masers in the Orion nebula have been studied in some detail, with observations showing them to move smoothly with time, covering distances that are many times their own diameters in periods of a year or two.

Figure 5.9 shows the speed and direction of some water-maser

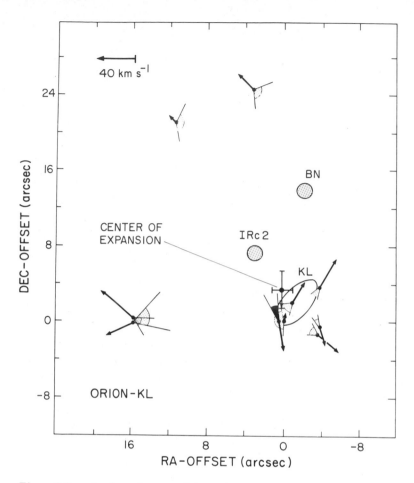

Figure 5.9: *A schematic map of the water-vapor masers and infrared stars (BN and IRc 2) in the Orion molecular cloud. The speed and direction of the motions (on the plane of the sky) of the masers are indicated with arrows. The masers appear to be expanding outward from a point near the very energetic, but highly obscured, new star IRc 2. (Smithsonian Astrophysical Observatory illustration by Joseph Singarella)*

spots in the Orion region. The motions clearly indicate expansion from a central position near a strong infrared-emitting nebula. These observations, coupled with recent infrared observations, indicate that a hitherto undistinguished infrared source, IRc 2, is really a massive star emitting energy at a rate of about 100,000 times that of the Sun. (IRc 2 stands for *Infra*Red *c*ompact source number *2*.) Not only is IRc 2 the primary energy source in the region, it is losing mass at a rate roughly one billion times more rapidly than the Sun.

Halfway across the Milky Way, another water-maser source, W51, has proved a valuable tool in the pursuit of still another astronomical goal: the determination of precise distances across the vastness of space. The water masers in this region have been mapped and motions observed, but, in contrast to the Orion source, no organized motion has been found. Instead, the water-maser spots appear to move randomly, suggesting that mass outflows in this region interact strongly with the surrounding molecular cloud to produce turbulence.

This random motion actually turns out to be a boon for astronomers, because its statistical properties can be used to determine distances to the star-forming regions. One obtains radial velocities (the velocity of an object either toward or away from an observer, in units of kilometers per second) from measured redshifts in the source spectrum. One then measures the masers' *angular motions* on the plane of the sky (in units of arcseconds/year). To convert the observed angular motions into linear velocities, one must multiply by the (unknown) distance to the source. Provided the motions are truly random, the spread of radial velocities should be equal to the spread of angular motions scaled by the distance. Therefore, by trying different values for the source distance until the spread in the motions on the plane of the sky matches the spread of the radial motions, one can determine, with some precision, a distance to the source. This technique is called "statistical parallax."

The distance to W51 determined in this manner is about 7,000 parsecs (1 parsec is about 3 light-years), a distance about 20 times greater than could be directly measured with existing optical techniques. Studies of this and other water masers have demonstrated that, with a sufficiently large sample of maser spots, the distance to a source can be determined with an accuracy of 10 to 20 percent anywhere in the Milky Way.

In addition to sources in our Milky Way, water masers have been detected in nearby spiral galaxies. Future observations with very sensitive arrays of radio telescopes should be able to map their motions also. Unfortunately, at a distance of 1,000,000 parsecs, water-maser spots moving at 30 kilometers per second (about the speed of the Earth around the Sun) would have angular motions of only 5 millionths of an arcsecond per year. While measuring so small a change in angle is challenging, such measurements do appear possible. Certainly, a *direct measurement* of the distance to another galaxy would be exceedingly important to our understanding of the size and age of the universe.

The Future

Early in this decade, a select committee of American astronomers issued a report to the National Academy of Sciences and the National Research Council outlining a plan for the orderly development of astronomy for the 1980s. The committee, chaired by George Field (see the final chapter), recommended that the highest priority new ground-based instruments be an array of radio telescopes spanning the United States east to west from Hawaii to New England and Puerto Rico and north to south from the state of Washington to southern New Mexico and Texas. The National Science Foundation has accepted this recommendation, and now astronomers eagerly await construction later this decade of the very long baseline array (VLBA).

Such a large array of radio telescopes would dramatically improve VLBI capabilities. In particular, the VLBA would roughly double the number of telescopes currently used for VLBI in the United States (Figure 5.10). Instead of relying on a handful of aging telescopes of varying quality located randomly, the VLBA would be designed with modern, high-performance telescopes carefully located so as to optimize the image-forming capability. Moreover, the telescopes would be dedicated solely to the task of interferometry; the current system is a jerry-rigged arrangement by which short blocks of time on the individual telescopes, each operated by a different institution, must be requested, allotted, and coordinated months in advance. The VLBA telescopes would reach wavelengths as short as 7 millimeters (and perhaps 3 millimeters), an improvement of about 5 (or 10) times over many of the present telescopes. This is particularly important for spectral line studies of the strong maser emissions of water vapor (1.3 centimeters) and silicon monoxide (3.5 and 7 millimeters). The

Figure 5.10: *Locations of the 10 elements of the new VLBA telescope: an "antenna the size of the Earth."* (Illustration from the National Radio Astronomy Observatory)

VLBA would have baselines (i.e., telescope separations) up to 8,000 kilometers, or nearly 3 times longer than currently used in the United States. This would produce nearly 10 times more picture elements per source. When used with compatible antennas in Europe, the VLBA system will represent a telescope with an aperture equivalent to the diameter of the Earth.

Equally important to extending baselines, the VLBA will produce images with much higher contrast. This will allow astronomers to see fainter emissions in the presence of strong source components. Such improvements will probably lead to greater understanding of the energy sources of radio galaxies and quasars.

The VLBA will also open up many new areas of galactic study, such as the study of red giant stars (including the long-period variable, or Mira-type, stars). These stars are in the process of dying, just as the Sun will do in some 5 billion years. They have grown in size by a factor of about 100 and are now shedding their outer atmospheres at an enormous rate. Embedded in the outer

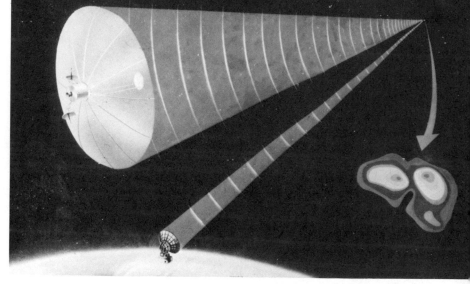

Figure 5.11: *A figure depicting the concept of an Earth-orbiting VLBI station. Space-based VLBI missions are being planned by NASA and the European Space Agency.* (NASA/Jet Propulsion Laboratory illustration)

atmospheres of the red giants are intense masers from several molecular species. By studying these stars with the VLBA, astronomers hope to learn how such stars lose their mass and how this ejected material is recycled within the Milky Way to form new stars.

Another important improvement of the VLBA is in the area of data analysis. The VLBA analysis system will be about 50 times more powerful than existing systems. It is being carefully designed for spectral line applications and for full polarization capabilities. Many types of observations currently difficult or impossible will be achieved because of these developments.

Once the VLBA is built, astronomers will be able to routinely make high-quality images in radio "light" with a resolution approaching one ten-thousandth of an arcsecond! Yet, even with this astounding capability, some of the most interesting questions of modern astrophysics will remain unanswered because of limited resolution. Better angular resolution than the VLBA can only be achieved by going to shorter wavelengths—or longer baselines. Shorter wavelength observations are severely limited by the absorption of the signals in the Earth's atmosphere and by the problem of building large telescopes with high-precision surfaces. On the other hand, to extend the distance between telescopes, one is forced to leave the Earth's surface.

The concept of putting a radio telescope in space is already being pursued in both the United States and Europe (Figure 5.11). A study of such a project, called Quasat for *Quasar Satellite*, is being carried out jointly by NASA and the European Space Agency (ESA). Current plans call for the deployment early in the 1990s of a small (about 15-meter diameter) radio telescope in an elliptical orbit reaching an altitude of about 20,000 to perhaps 50,000 kilometers. This would at least triple the lengths of baseline available with the VLBA. Using Quasat with the VLBA would make it possible to construct images with at least 10 times more picture elements than with the VLBA alone.

By combining all of these improvements, the VLBA and Quasat should revolutionize radio interferometric observations. Astronomers look forward with great anticipation to studying the many celestial objects available to the new instruments. With the VLBA and Quasat, we should be able to observe both ends of the star formation cycle: from newborn stars that form out of the interstellar material to the dying red giants that largely resupply the interstellar medium through stellar mass loss. We can study luminous masers and "dark" absorbing hydrogen gas, and determine the three-dimensional magnetic field vectors and the three-dimensional motions of gas from many classes of objects. We can probe deeply into the very hearts of radio galaxies and quasars, perhaps at last revealing the nature of their energy sources, compared to which even nuclear energy seems feeble. And, finally, if the pattern of discovery following new instruments continues with the VLBA and Quasat, we can look forward to the prospect of finding a great variety of unknown—even unimagined—objects and phenomena.

Shortwave Radio

Millimeter and Submillimeter Astronomy

PAUL T. P. HO

During the past decade, some of the most exciting progress in radio astronomy has been made in the relatively narrow millimeter and submillimeter wavelength windows, a still largely unexplored region of the electromagnetic spectrum between the infrared band and the longer wavelengths of the radio band. This has been brought about primarily by the development of improved millimeter receivers, where sensitivity has increased by roughly a factor of 400. Moreover, the successful application of heterodyne techniques (that is, combining signals of different frequencies to produce "beats") to the millimeter–submillimeter band has led to many important discoveries: spectral lines of some 64 molecules, giant molecular clouds (GMC) that are apparently the most massive components of the Milky Way, large-scale structures both in our own galaxy and others, and exotic phenomena such as massive outflows from young stars.

The millimeter–submillimeter wavelength band is important because the bulk of the interstellar medium is cool, with a temperature of approximately 10 degrees Kelvin,* and can be sampled only by radio astronomical techniques. Important phenomena associated with the star-formation processes, as well as the organization of the gaseous constituents of the galaxy into large-scale structures, are best studied at these wavelengths. Moreover, dust-continuum emission becomes detectable in the submillimeter wavelengths, and there is an abundance of molecular species useful for sampling a wide regime of different physical conditions. By choosing the appropriate molecular transition, it is possible to study any part of the interstellar medium. With heterodyne techniques, the spectral lines of these molecules can be resolved; and, by measuring accurately the Doppler shifts in line frequency, motions can be easily traced. Determining the dynamics is most important, for it allows us to infer stability, lifetimes, energetics, and hence the eventual evolution of the interstellar medium.

Major technological advances have made these studies possible. In addition to greatly improved receivers, construction is completed or under way on a number of large millimeter-wave and submillimeter-wave telescopes with apertures in the 30-meter class. The increased collecting area means both increased sensitivity and angular resolution. In a parallel vein, interferometric techniques are being successfully applied in the millimeter-wave range. The synthesis of a very large aperture—in the 300-meter region—by using widely separated antennas has brought about arcsecond angular resolution. The large magnification provided by interferometry is most exciting because of the detailed information it provides on a variety of phenomena ranging from molecular cloud cores to the dynamics of galactic nuclei.

In this chapter, I review some recent and exciting results in the field of millimeter- and submillimeter-wave astronomy. I describe the premier instruments currently available or under construction, and the prospects for discovery promised by the next generation of instruments.

Major Discoveries

It is impossible to give a comprehensive summary of all the important results of the past 15 years. Rather, I will highlight some

---

* A Kelvin (K) is a unit of temperature the same size as a Celsius degree but with the zero point set at absolute zero, that is, $-459.4°$ F, or $-273.2°$ C. Thus, a typical room temperature can be expressed as 68° F, 15° C, or 288° K.

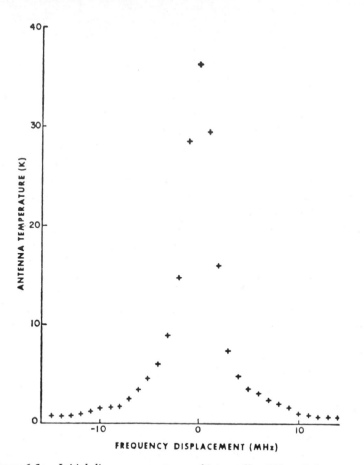

Figure 6.1: *Initial discovery spectrum of interstellar CO emission toward the Orion nebula in 1970 by Wilson, Penzias, and Jefferts. Note the high brightness temperature of the emission and the non-Gaussian nature of the line wings. Even in this initial study, it was determined that CO emission is spatially extended and widely detectable.* (Illustration courtesy of Robert Wilson)

discoveries that, in my opinion, have been important not only in their own right but also for their ramifications on the rest of astronomy.

Molecules. We now know that the bulk of the interstellar medium consists of hydrogen, either in atomic form, denoted as HI, or in molecular form, denoted as H_2. HI can be detected directly because its spin-flip produces a clear radio signal with a wavelength

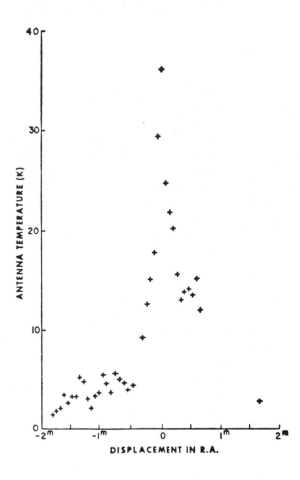

of 21 centimeters. However, because H_2 is a symmetrical molecule, this form of hydrogen is extremely difficult to excite and detect directly in the ordinary interstellar medium. Thus, the discovery of other molecules in space, each with a distinctive millimeter-wave signal, has proved to be a great asset because the much less abundant molecular species can be used to trace the predominant H_2 distributions.

At the last count, some 64 molecules had been found in the interstellar medium, most of them detected in the millimeter-wave range. Of these, the most important is clearly carbon monoxide (CO), first detected in 1970 by Bob Wilson, Ken Jefferts, and Arno Penzias of Bell Laboratories, using the 11-meter telescope of the National Radio Astronomy Observatory (NRAO). Their original detection in the direction of the Orion nebula is shown in Figure 6.1. In this early experiment, they established that the CO line radiation was, as astronomers term it, bright, broad in velocity profile, and extended over the sky.

Because the CO line is so intense and widely distributed in space, this molecule is a most useful tool for tracing the interstellar medium. In addition, measurements of the rarer isotopes of this molecule have shown that the main isotopic line $^{12}C^{16}O$ is optically thick, that is, the column density of the molecule is so high that the material becomes opaque at the transition frequency. Moreover, the upper-energy levels of the CO molecule are easily excited by collisions with hydrogen molecules. The combination of high optical depth and ease of excitation implies that CO emission brightness will accurately reflect the local gas temperature. With improved receiver systems, it has been shown that the CO molecule is an accurate thermometer for most molecular material. Furthermore, CO is the most abundant molecular species by far, and it is essentially ubiquitous, found everywhere in the galaxy where densities exceed 10^2 H_2 molecules per cubic centimeter.

Numerous other molecules, some in minute, or "trace," quantities, have been found in the interstellar medium. These trace constituents have very different excitation requirements. Thus, by picking an appropriate molecular transition, it is possible to probe a specific layer within these gaseous complexes known as molecular clouds. Because a star-forming cloud complex is an onionlike structure, with succeeding layers having higher densities and temperatures, it is possible to measure a density, mass, and temperature profile by using a variety of molecular tracers. With so many molecules detected, and their abundances measured, it is now possible to consider seriously interstellar chemistry. In fact, the detection of important ions such as HCO^+ has led to the now commonly held belief that ion–molecule reactions are important in the interstellar medium. The abundance of ions can also be used for determining the state of ionization and for inferring the abundance of electrons. Thus, ions not only determine the chemistry, but also serve to retain the magnetic fields within cloud complexes.

One final point about molecular lines: they are coolants for the interstellar medium. The intense CO radiation serves as an effective means for isolated gas complexes to lose energy to their surroundings. In the balance of heating and cooling of the interstellar

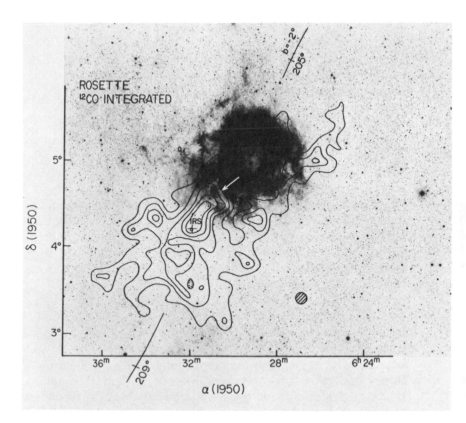

Figure 6.2: *The giant molecular cloud detected toward the Rosette nebula by Blitz and Thaddeus. The contours show the intensity of CO emission, superposed on an optical image of the nebula. The molecular complex is extended along the galactic plane with a sizescale of about 100 parsecs.* (Illustration courtesy of Leo Blitz)

medium, line radiation by the CO molecules, as well as other species, plays an important role.

Giant Molecular Clouds. One of the original conclusions from the initial detection of CO was that the emission is widely distributed on the sky. Subsequent CO studies revealed a heretofore unknown component of the galaxy—giant molecular clouds (GMC). These giant gaseous complexes have sizescales on the order of 50 to 100 parsecs and have masses on the order of 10^5 to 10^6 times that of the Sun (Figure 6.2). The complexes are also cool, with temperatures of approximately 10 degrees Kelvin, and dense, with an average density of 300 H_2 molecules per cubic centimeter. Because of their typical distances, and because of intervening dust along the line of sight, the GMCs cannot be seen visually. However, because dust extinction is negligible at radio wavelengths, the GMCs are clearly delineated by their CO emission lines.

These giant complexes are important simply because they are numerous. Surveys of the CO emission in the Milky Way suggest some 4,000 GMCs are distributed across the galactic plane in a layer about 100 parsec thick. With a total mass of approximately 10 billion Suns, these cloud complexes are the dominant component of the interstellar medium. The GMCs are themselves quite clumpy, with a number of dense cores; and some cores are active sites of star formation. In fact, all young stars are found to be forming at some dense spot within a GMC. Hence, the GMCs can be identified as the current—and future—sites of star birth.

Some important questions concerning these giant complexes remain unanswered. First, the exact spatial distribution of the GMCs in the galactic plane is controversial. Are GMCs found only in the spiral arms, or are they distributed throughout the disk? This is a key issue, for it concerns the origin, formation, and lifetimes of these clouds. Are they compressed versions of giant HI complexes that have passed through the spiral arms, or does star formation caused by the spiral arms simply light up the molecular complexes? A second crucial question concerns the future evolution of the GMCs. Current theories suggest that star-formation processes, such as the rapid expansion of the ionized gases around massive stars, as well as the rare supernova event, will eventually disrupt and destroy the surrounding molecular cloud. Will this material be completely dispersed, or is it only temporarily "inflated"? Whatever the answers to these questions, it is clear that all new stars are being born within the giant molecular clouds.

Galactic Structures. As the most massive component of the interstellar medium, GMCs obviously play an important role in galactic dynamics. Spectroscopy provides a key to that role, since by virtue of the Doppler shifts in line frequency, relative motions can be measured. With this velocity information, the GMCs can be studied in the framework of a rotating galaxy, with a three-dimensional picture constructed in the context of a dynamical model. Once the spatial distribution is determined, one can examine the molecular complexes for evidence of cohesive structures, such as alignment along the spiral arms.

Using a 1-meter millimeter-wave telescope on top of the physics building at Columbia University, a group led by Pat Thaddeus has surveyed completely the northern part of the galactic plane. By comparing with similar HI surveys, the Columbia group concluded that the molecular material delineated the same spiral arms first identified through HI studies. The same conclusion is reached by a separate group led by Phil Solomon and Nick Scoville, who made a similar survey with the 11-meter NRAO telescope and the 14-meter telescope of the Five College Radio Astronomy Observatory (FCRAO) (Figure 6.3).

After determining the three-dimensional distribution of the molecular material, one can also examine its galactocentric distribution, that is, its distance from the center of the galaxy. In fact, the molecular material seems to be concentrated in a ring between 4 and 8 kiloparsec from the galactic center. The exact cause of this ringlike distribution is unclear. It could be due to explosive events in the nuclear region, or perhaps to resonances associated with the spiral arm structures.

Because the identification of spiral features is difficult from within our own galaxy, the best hope of understanding the spatial distribution of GMCs may be in observing other galaxies. Since the first discovery of extragalactic CO in 1975, the measurements of CO emission in external galaxies have steadily become more and more important for millimeter-wave astronomy. Already some 100 galaxies have been detected in CO, and careful mapping of some nearby galaxies has begun. Using the 7-meter Bell Laboratories telescope, Tony Stark and collaborators mapped a sector of the Andromeda galaxy (M31) and found that the CO emission is strongly concentrated in its spiral arms (Figure 6.4). On the other hand, the interstellar medium of M31 is predominantly atomic, with less than 10 percent of the hydrogen in molecular form. This is in sharp contrast to our own galaxy, where the interstellar medium is predominantly molecular. More detailed studies with higher angular resolution are needed.

Massive Outflows from Young Stars.   Even in the initial detection of CO emission toward Orion, there were clear indications of some unusual and spectacular events occurring within the molecular cloud.

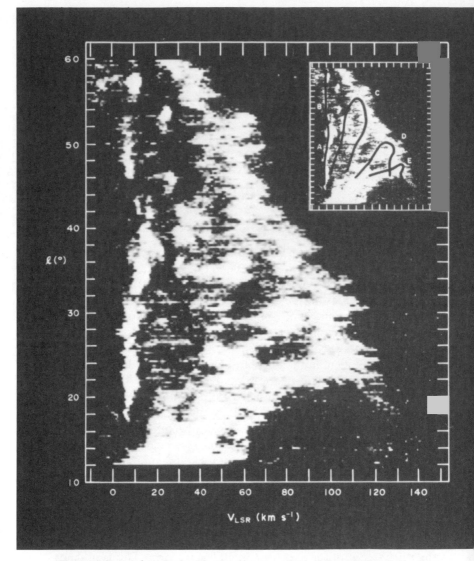

Figure 6.3: *A longitude-velocity diagram of the CO emission in the first quadrant of the galaxy, from Cohen, Cong, Dame, and Thaddeus. This diagram can be most easily interpreted as a stack of spectra with the ordinate axis being the displacement along the galactic plane away from the galactic center. The perceived motions as one moves away from the center can be interpreted in terms of spiral arms. Proposed spiral arms from HI studies are indicated in the inset. The CO emission can be seen to reside mostly in spiral arms.* (Photograph courtesy of Richard Cohen)

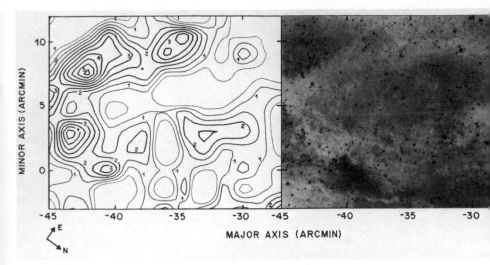

Figure 6.4: *CO emission in the nearby Andromeda galaxy, as reported by Stark. The contours of CO intensity (left) are seen to correspond very well with extinction in an optical image (right) of the same field. This demonstrates the excellent correlation between dust and gas as traced by CO. This also shows that giant molecular clouds lie predominantly in the spiral arms of galaxies.* (Photograph and illustration courtesy of Anthony Stark)

The high-velocity line wings appearing in the data (see Figure 6.1) represented substantial excess emission far above any well-behaved statistical distribution in velocity. The phenomenon has since been interpreted as real motions of large magnitude. (More recent high-sensitivity spectra show evidence of motions up to ± 100 kilometers per second.) Given the magnitude and the spatial sizescale of this motion, the amount of mass required to gravitationally bind it would be enormous—the equivalent of 4,000 solar masses, or many times more than all the mass in the core of the Orion cloud. Thus, the motions must be unbounded and probably represent an outflow of material from a central star.

Although the idea of mass outflows from young stars was intriguing, it remained dormant until 1980, when two new sources of outflows were discovered, one in the dark cloud Lynds 1551 and another in the star-formation molecular complex in Cepheus. These new outflows were of great interest because they appeared to be anisotropic, that is, displaying different physical properties in dif-

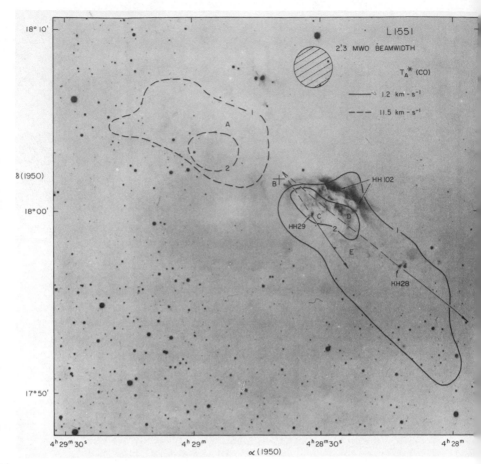

(a)

Figure 6.5: *(a) Outflow of high-velocity material in the nearby dark cloud
L1551, as reported by Snell, Loren, and Plambeck. The solid contour
indicates blueshifted (moving toward us) material, while the dashed contour
indicates redshifted (moving away from us) material. The high-velocity
material is seen to be ejected in opposite directions and has a well-collimated
appearance. This bipolar morphology has now been found in a number of
other regions. (b) The schematic indicates a likely model of the observations.*
(Photograph and illustration courtesy of Ronald Snell)

ferent directions. By comparing the morphology of that portion
of the outflow moving away from us (redshifted) to the portion
moving toward us (blueshifted), it was clear that the two portions

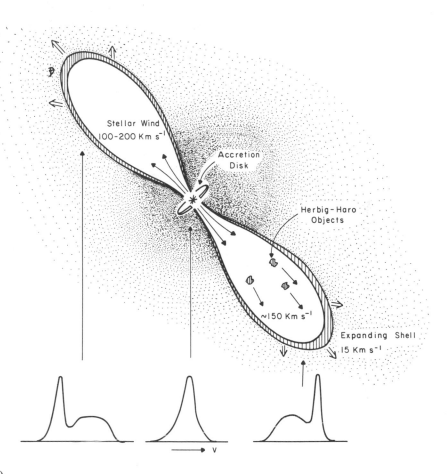

(b)

are displaced in the sky. The morphology suggests that the outflows may be collimated into two distinct lobes, a phenomenon now known as *bipolar outflow* (Figure 6.5).

Many more outflows have since been found in the interstellar medium. However, it remains unclear how many of the outflows are anisotropic. Furthermore, the mechanism responsible for collimating the outflows remains unknown. What is clear is that the outflow phenomenon must be common in the lifetime of a young star. Even in the rather incomplete survey of the solar neighborhood, the large number of detected outflow sources, combined with the apparent lifetime during which an outflow can be recognized, implies a high formation rate. Because outflows are typically found in dense clouds centered on an infrared source, or a very young main-sequence star, it is almost certain that the phenomenon is associated with a newly born star, or even a protostellar object. Perhaps every star more massive than a few solar masses undergoes such an eruptive or outflow phase.

The importance of mass outflows is that they are apparently so common. The total mechanical energy deposited into the interstellar medium over time is substantial, perhaps second only to supernova events. Mass outflows may also serve to reverse the accretion process during the last stages of making a star. (This point is still unclear, since it is not known exactly when the outflows are turned on.) That this phenomenon has been recognized only recently as common and important is purely due to the sensitivity of our instruments. Much work lies ahead in defining the detailed morphology, locating possible focusing mechanisms, and understanding the actual outflow process from the surface of the star.

Millimeter-wave Interferometry.   During the early 1970s, in the relative infancy of millimeter-wave astronomy, efforts were already under way to build interferometers that might achieve angular resolutions of 1 to 2 arcseconds. The earliest instrument was a two-element interferometer constructed by the observatory in Bordeaux, France, operating at 8 millimeters. This interferometer, conceived in the late 1960s by Jean Delannoy, Jacques Lacroix, and Emile Blum, was designed for solar studies and made use of two 2.5-meter antennas. A second group at the University of California at Berkeley led by Jack Welch built a two-element movable interferometer with bigger antennas (6 meters) and designed to study the then newly discovered millimeter-wave molecular lines. By the late 1970s, receiver technology had improved considerably, and one of the early results from Berkeley was a map of the core of the Orion molecular cloud in the sulfur monoxide (SO) molecular line, at a wavelength of 3.4 millimeters (Figure 6.6). A small condensation, about 15 arcseconds in size, was found in the core of the Kleinmann-Low infrared nebula. This SO condensation appears coincident with the embedded infrared sources and so well-centered with respect to the observed CO outflow in the region that it may be interacting with the outflow or even serving as the focusing agent for the outflow. As the Berkeley maps became increasingly more detailed, the SO condensation was found to be an elliptical, elongated structure, well centered on a bright infrared

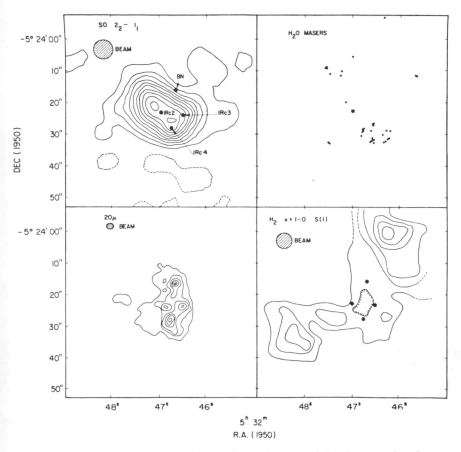

Figure 6.6: *A composite of activities in the core of the Orion molecular cloud. The SO emission is mapped with 6-arcsecond resolution, using the Berkeley millimeter-wave interferometer. The SO emission demonstrates the presence of a dense gaseous core closely associated with the infrared object IRc 2. The H_2O masers represent a high-velocity outflow, while the vibrationally excited H_2 represents highly excited, or shocked, gas. Other compact infrared objects are present as indicated by the 20-micron emission.*

source and lying nearly perpendicular to two lobes of H_2 infrared emission lines representative of a highly excited, probably shocked, gas. The orientation of the SO condensation suggests a disk around the exciting source. The SO's measured brightness temperature of 50 degrees Kelvin also suggested a high density, and the measured kinematics indicated relative motions between the two opposite edges of the disk. Thus, a possible model for this source is an expanding "doughnut" of gas where an energetic outflow is pouring out of the hole in the center. These results suggest that a wealth of other details awaits discovery by instruments with arcsecond-class resolution.

During the early 1980s, the California Institute of Technology also constructed a millimeter-wave interferometer consisting of two 10-meter antennas in Owens Valley, California. One of the first Caltech results was a map of the nuclear region of the nearby spiral galaxy IC342 (Figure 6.7). With 7-arcsecond resolution, the map revealed that the molecular material is distributed in a bar that appears to be rotating. The presence of this bar, which is not seen in the optical wavelengths, suggests the presence of an oval distortion in the nuclear gravitational field. Possibly an inward flow of gas induced by such a perturbation in the gravitational field may enhance star-formation activities in the nucleus. Again, high angular resolution revealed structure and dynamics previously unsuspected.

Both the Berkeley and Caltech interferometers have recently added third elements. And other interferometers now being built promise arcsecond-type resolution that will surely provide many new details of cloud structures and dynamics.

Submillimeter Astronomy.   The submillimeter band encompasses wavelengths between 1 and 0.1 millimeter. In this window, atmospheric extinction is quite severe due principally to the absorption bands of water vapor. Ideally, then, the most successful observations should be made from above the atmosphere on a space platform. Since satellite-borne telescopes for this wavelength band are not yet available, astronomers are currently using two alternatives: high-flying aircraft (e.g., NASA's Kuiper Airborne Observatory) and high mountain sites (e.g., Mauna Kea at 4,200-meter elevation).

There are a number of important submillimeter spectral lines. For example, the lines of metal hydrides, if detected, will give the abundances of these metals. The higher rotational lines of such molecules as CO and HCN can sample gases with temperatures greater than 100 degrees Kelvin. And some important species, such as atomic carbon (CI), are seen in the submillimeter band. Finally, at submillimeter wavelengths, continuum emission from dust be-

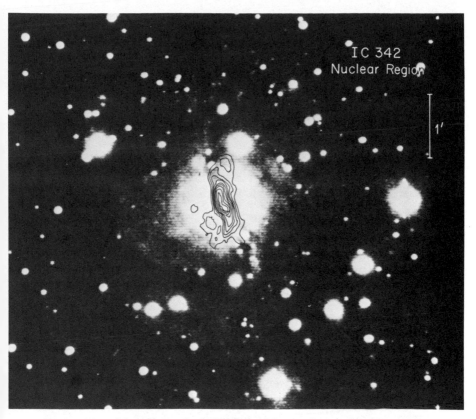

Figure 6.7: *CO emission in the nucleus of the nearby spiral galaxy IC342. This high-resolution map is obtained with the Caltech millimeter-wave interferometer. The presence of a barlike structure may be related to gas flows into the nuclear region, providing raw material to support strong bursts of star formation.* (Photograph courtesy of Kwok-Yung Lo)

comes important. For most of the interstellar medium, the dust continuum approaches a blackbody spectrum. At typical dust temperatures of 20–100 degrees Kelvin, the peak of the dust emission is in the far infrared, with wavelengths shorter than 0.1 millimeter. However, some portion of this dust begins to be measurable in submillimeter wavelengths.

Receiver systems in the submillimeter band are still fairly insensitive. Nevertheless, significant discoveries are already being made. For example, emission from the higher rotational levels of

CO has been detected in the Orion molecular cloud, as has the atomic carbon line. (The atomic carbon measurement is very important because, by comparing CI to CO abundances, it is possible to estimate the percentage of carbon residing in the form of carbon monoxide.) A University of Chicago group led by Roger Hildebrand has been studying dust emission surrounding very young stars and also in external galaxies. And dust continuum emission in the submillimeter wavelengths has even been detected recently in a compact dark cloud. Together with measurements in the far infrared, dust temperatures and luminosities can be estimated with good precision.

Two exciting problems in astrophysics concern the emission mechanisms in quasars and active galactic nuclei, and the microwave continuum background. In both cases, the submillimeter wavelength band may hold important keys to their solution.

For quasars and active galactic nuclei, astronomers suspect that a compact nuclear core serves as the central engine responsible for the enormous outpouring of energy. The electromagnetic spectrum for the core source typically rises in the infrared toward the submillimeter band, turns over, and heads down again in the radio band. The exact wavelength at which the spectrum turns over can be used to distinguish between the various emission mechanisms that have been proposed as "drivers," such as synchrotron sources and relativistic jets. Since the submillimeter band is apparently where the spectrum turns over, measurements of a large number of these sources will be crucial for our understanding of the central engines. The construction of large submillimeter-wave telescopes will make these measurements possible.

Since the discovery of the microwave background radiation associated with the Big Bang, it has become recognized that Compton scattering by the hot intergalactic gas in clusters of galaxies may alter the cosmic blackbody spectrum. The so-called Sunyaev-Zeldovich effect is expected to produce a decrement at a wavelength of 3 millimeters and an excess at 0.8 millimeter. The magnitude and angular size of this effect, when combined with estimates of the electron density and temperature from X-ray measurements, can yield a distance estimate for the clusters of galaxies. Such distance measurements, coupled with a measurement of the velocity of the cluster, will yield an estimate of the Hubble constant. In other words, sensitive measurements at millimeter and submillimeter wavelengths may be vital to a determination of the age of the universe.

New Telescopes

Several major efforts are under way to construct new instruments with larger collecting areas and higher angular resolutions for millimeter and submillimeter wavelengths. Several technical problems that have hindered advances in this field appear now to be solvable.

First, at these wavelengths, the surface, or shape, of the antenna must be precisely figured. In order for scattering losses to be small, the final surface should be accurate to about 5 percent of the operating wavelength. Hence, a good surface for submillimeter astronomy must be accurate to approximately 20 microns. Fortunately, this kind of accuracy can now be achieved with computer-controlled fabrication techniques using a variety of materials, including aluminum panels, glass, or carbon-fiber-reinforced plastics.

Second, because the telescopes that are being built are quite large, typically with diameters 30,000 times the operating wavelength, the primary diffraction-limited beam size, or field of view, is approximately 10 arcseconds. This means the telescope must point very well—at the arcsecond level! Because thermal instabilities can both distort the surface and affect the pointing, telescope makers must either insulate and/or actively control the temperature of the instrument to a few tenths of a degree, or use carbon-fiber material, which has a very small coefficient of thermal expansion.

Third, to measure and set the antenna surface accurately, two standard approaches have been developed. Laser ranging to corner reflectors set on individual panels of the surface has been used. Alternately, it is possible to make holographic tests using the interference pattern (in conjunction with a second telescope) across the aperture of the telescope.

Finally, the interferometers operated by the University of California at Berkeley and Caltech have shown that aperture synthesis techniques can work successfully in the millimeter-wave range, providing angular resolution on the order of 1 arcsecond.

Single-element Telescopes. Almost all the early millimeter-wave spectroscopy was done with the 11-meter telescope at Kitt Peak operated by the NRAO, and the 5-meter telescope of the University of Texas at Austin. In the late 1970s, two larger millimeter-wave telescopes were built: the 14-meter telescope of the FCRAO in Amherst, Massachusetts, and the 20-meter telescope of the Onsala (Sweden) Space Observatory. A number of smaller instruments were also in operation during that time, including the 1-meter telescope of Columbia University, the 5-meter telescope of the Aerospace Corporation, and the 7-meter telescope of Bell Laboratories. All these telescopes were designed for a wavelength of 3 millimeters, although some have now been pushed to a wavelength as short as 1 millimeter.

In the late 1970s, plans were made to build much larger millimeter-wave telescopes. An American effort to build a 25-meter telescope was not successful, but two other efforts did succeed. The University of Tokyo constructed a 45-meter open-air telescope in Nobeyama, Japan, and a French-German collaboration, the Institut de Radio Astronomie Millimetrique (IRAM), has just completed the construction of a 30-meter open-air telescope at Pico Velata near Granada, Spain. These two large millimeter-wave telescopes are truly in a class by themselves. At their respective shortest operating wavelength, they are unique in terms of angular resolution and sensitivity, and should contribute much to extragalactic studies (Figure 6.8).

In submillimeter astronomy, there are as yet no instruments dedicated exclusively to this wavelength band. To date, all the astronomical measurements through this window have been made on infrared or optical telescopes. However, several major efforts are now under way to place submillimeter instruments at good dry sites. For example, Caltech is placing a 10-meter telescope on Mauna Kea in Hawaii. A British-Dutch collaboration is also building a submillimeter-wave telescope on Mauna Kea, adjacent to the Caltech instrument. The plans call for a 15-meter telescope with aluminum panels accurate to 15 microns. Its surface will be roughly two times less accurate than the Caltech surface; however, its collecting area is about twice as great. Between 1 and 0.5 millimeter, this telescope will have an angular resolution of 10–20 arcseconds. An American-German collaboration (University of Arizona/Max Planck Institut fur Radioastronomie) is also building a 10-meter submillimeter-wave telescope on Mt. Graham in Arizona (Figure 6.9). This telescope will make use of the carbon-fiber technology both for the skins of the aluminum honeycomb panels

Figure 6.8: *The 30-meter IRAM telescope at Pico Velata, Spain.* (Photograph courtesy of Dennis Downes)

and the backup structure. Its performances should be equivalent to the Caltech 10-meter telescope. All three telescopes will reside in co-rotating insulated buildings with either barn-type or rollback doors.

Two more European efforts in single-dish submillimeter-wave astronomy include the University of Cologne's existing 3-meter telescope for millimeter astronomy, which will be moved to the Swiss Alps where the higher altitude will allow submillimeter studies. Also the Onsala Space Observatory and the European Space Agency intend to construct a 15-meter telescope at the European Southern Observatory in LaSilla, Chile. The telescope will surely utilize carbon-fiber technology, but detailed construction plans are not yet available.

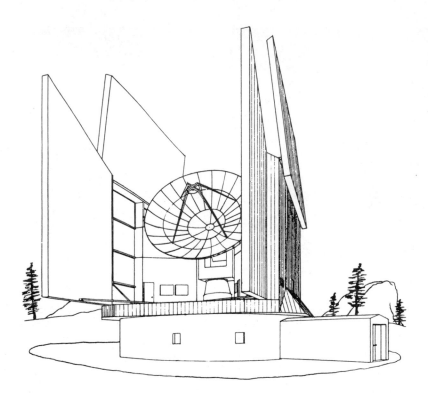

Figure 6.9: *Artist's rendition of the 10-meter submillimeter-wave telescope being built by the University of Arizona and Max Planck Institut.* (University of Arizona/Steward Observatory illustration)

There is a great deal of international effort invested in single-element telescopes for both millimeter- and submillimeter-wave astronomy. The plans all call for about 10-arcsecond angular resolution, which represents a factor of 5 or 6 improvement over existing single-element telescopes.

Multiple-element Interferometers. To provide angular resolutions substantially better than 10 arcseconds, one must resort to interferometry. The single-element telescope approach is impractical because it would be extremely difficult and expensive to build a very large telescope and maintain its shape against thermal and gravitational deformations. Moreover, the extreme precision with which one must be able to point such a telescope also seems beyond present technical capabilities. The "aperture synthesis" approach of simulating a large aperture by using a number of smaller telescopes has a number of advantages. First, since the signal from each telescope is available before the formation of the interference pattern, it is possible to map the entire primary beam by inserting the appropriate phase delay for each telescope. Second, by image restoration techniques, it is possible to correct for the imperfect sampling of the aperture. Third, since individual telescopes are small, the primary beam, or field of view, is large. (Furthermore, both pointing and surface accuracies of the telescopes are more manageable.) Fourth, the angular resolution of the interferometer is ultimately limited only by the phase stability, or clarity, of the atmosphere above the telescope.

Both the Berkeley and Caltech millimeter-wave interferometers have already demonstrated the feasibility of aperture synthesis at the millimeter wavelengths, and both instruments are now in the process of being upgraded (Figures 6.10 and 6.11).

Two more millimeter-wave interferometers will soon become operational. The Japanese group at Nobeyama has constructed an array of five 10-meter telescopes located at the same site as its 45-meter telescope. If the array and the single-dish telescope are operated together as an interferometer, the resulting instrument will be powerful, fast, and sensitive. The French-German Institute (IRAM) is constructing an interferometer of three 15-meter telescopes on the Plateau de Beure near Grenoble, France. Although this interferometer is not as fast as the Japanese system, it has almost the same collecting area. With its better surfaces and its higher altitude, the IRAM instrument may achieve operations down to a wavelength of 1 millimeter.

An American effort, led by the NRAO, hopes to construct a national millimeter-wave interferometer. The initial design calls for some 1,000 square meters of collecting area, at least 20 telescopes to ensure a fast instrument, location at a high, dry site, and operations down to 1-millimeter wavelength. A study group has been formed to consider possible configurations, size of telescopes, possible sites, and technical requirements.

In the submillimeter range, another American effort, led by the Smithsonian Astrophysical Observatory, proposes to construct a pioneering instrument that will have an array of six 6-meter telescopes located at a high site with as little water content in the

Figure 6.10: *The three-element interferometer of Caltech at Owens Valley, California.* (Photograph courtesy of Kwok-Yung Lo)

atmosphere as possible, and that will operate down to 0.35 millimeter. Both Mauna Kea and Mt. Graham seem to be promising sites, especially because of the presence of single-element submillimeter telescopes on these mountains. Linking the interferometer to other telescopes is an important factor because of the increase in collecting area. It is also crucial that the submillimeter array be a fast instrument, since weather variations on the scale of a week can affect submillimeter observations. Indeed, synthesis without reconfiguration of the array would be a definite advantage. The design of this instrument is still at an early stage. However, the prospect of studying dust emission with arcsecond-type reso-

lution in the galaxy as well as extragalactic sources is most exciting
(Figure 6.12)

Spaceborne Platforms. At millimeter wavelengths, the atmosphere is sufficiently transparent so that an instrument in space is
probably not warranted. In fact, a very large array, such as the
multi-element system envisioned by the NRAO group, will probably be the forefront instrument for many years. (This is similar
to the case at centimeter wavelengths where the Very Large Array
in New Mexico will be the state-of-the-art instrument for some
time.) At submillimeter wavelengths, however, much can be gained

Figure 6.11: *The millimeter-wave interferometer of the University of California, Berkeley.* (Photograph courtesy of William J. Welch)

by getting above the atmosphere. There are currently two proposals to launch permanent orbiting facilities.

The National Aeronautics and Space Administration has described plans to launch a 20-meter telescope known as the Large Deployable Reflector (LDR), which would be carried into orbit by the Space Shuttle. The initial design calls for diffraction-limited operation down to 30–50 microns. With a resolution of 1 arcsecond at 0.1 millimeter, this telescope could fill in all the gaps between 1 and 0.03 millimeter rendered opaque by the Earth's atmosphere, including some regions of the far infrared. The detailed design of this telescope has not yet been completed, and one alternative approach might be an array of smaller telescopes that operates as an interferometer (Figure 6.13).

Figure 6.12: *Artist's rendition of the Smithsonian's proposed submillimeter telescope array.* (Smithsonian Astrophysical Observatory illustration by Robert L. Marvin)

Figure 6.13: *Artist's rendition of the LDR being deployed by astronauts from the Shuttle.* (NASA illustration)

The European Space Agency is developing an 8-meter telescope called the Far Infrared and Submillimeter Space Telescope (FIRST). The design calls for operation between 1 and 0.1 millimeter. The beam size will be 3 arcseconds at 0.1 millimeter. Since FIRST will probably be launched by an Ariane IV rocket that has a maximum shroud diameter of 3.6 meters, the telescope needs to be folded for launch and deployed in space. In terms of scientific purpose and wavelength coverage, FIRST and LDR are very similar; however, because of its smaller size and somewhat longer operating wavelength, the surface tolerance of FIRST is less critical.

For the spaceborne telescopes, the main difficulties will be surface alignment in space and servicing of the various focal-plane photometers, spectrometers, and receivers on board. As most equipment will require cryogenics support, the ability to retrieve the satellites and recharge the cooling systems will be essential. This capability will also be useful for replacing on-board instruments with up-to-date equipment.

The prospects of doing submillimeter–far-infrared astronomy from space are most exciting. There is a wealth of phenomena whose radiation peaks in this wavelength band, and until now, most of this band has been invisible to us because of the atmosphere. These space telescopes will allow us to penetrate this barrier and at the same time achieve arcsecond angular resolution.

Prospects

A number of important astrophysical problems are waiting to be solved by the technical advances in millimeter and submillimeter astronomy. For example, the emission in submillimeter wavelengths of both the dust and the molecular constituents provides powerful tools for studying the higher temperature regimes of interstellar space. The high-temperature zones delineate the cores of molecular clouds where intense star-formation activities are taking place. Molecular probes, spectroscopic determinations of column densities and kinematics, and interferometric techniques to provide fine angular resolution and hence the magnification, form a most powerful combination for illuminating the basic processes and mechanisms important to the creation of stars. With improved sensitivities at both millimeter and submillimeter wavelengths, many important questions about galaxies may be answered, including the global distribution of the gas and dust, their kinematics, and their relation to star-formation activities. The overall relation between the atomic and molecular hydrogen distributions, the gas content in relation to morphological types, and the engines that drive the activities in the nuclei of galaxies and quasars can also be addressed. Finally, measurements of the microwave background radiation may have great cosmological implications. Small-scale anisotropy in the background radiation may reflect the density structures of the early universe, and the scattering and distortion of the background radiation by hot gas in clusters of galaxies may yield more precise distance estimates, and, in turn, insight on the deceleration of the expanding universe. The millimeter and sub-millimeter bands are the best windows for studying all of these questions.

That it is possible to contemplate—or even to imagine—solutions to these important questions must be accredited to the great technological advances being made in earthbound laboratories. The continued and rapid improvements in receiver sensitivity are allied with international efforts to construct superior instruments. Still, the large-aperture, single-element telescopes and the multiple-element interferometers are only the first steps in our effort to increase light-gathering ability and imaging capabilities. Future satellite experiments will carry aloft another generation of instruments to improve our vision in this wavelength band and to close the gap between the optical and microwave windows.

The Cool Sky

Infrared Astronomy

STEVEN P. WILLNER

Astronomers are just now beginning a thorough exploration of the infrared sky. Indeed, this window on the universe has been opened only a crack. A recent NASA catalog of all infrared observations made between 1965 and 1982 could list only some 10,000 objects. Considering the serious limitations imposed by detector technology and by the fact that most observations had to be made from the Earth's surface, the list is impressive. However, by comparison, the Henry Draper catalog of visible stars, published around the turn of the century, contains over 200,000 stars. Since the infrared region of the spectrum comprises a wavelength interval of 10 octaves compared to 1 octave for the visible, the small number of sources measured represents an even greater degree of ignorance than the bare numbers of sources at first suggest.

Fortunately, the technology for detecting infrared radiation is now rapidly improving. Within a few years, infrared detectors should be as good as those now available for visible light. Satellite observatories will offer the possibility of measuring much fainter objects at all wavelengths, not just the few wavelengths that penetrate the Earth's atmosphere. Indeed, the next decade should see an explosion in the number of infrared objects observed, and many of our current concepts about the universe are certain to be overturned.

In fact, the first giant step in the exploration of the infrared sky has already been taken. During most of 1983, the joint U.S.–Netherlands–UK Infrared Astronomical Satellite (IRAS) performed an all-sky survey that measured approximately 250,000 different celestial sources. The data, which were still being analyzed as this was written, have changed our understanding of many different types of objects. Among other things, we have already seen the strong emission both along the Milky Way and in regions where stars are forming, distant galaxies that are faint in visible light but among the most luminous infrared sources in the sky, and the presence of possible preplanetary material in orbit around Sun-like stars.

Future studies aimed at detailed investigations of particular classes of celestial objects will be conducted with a variety of observing instruments on the ground, on airplanes, balloons, or rockets, and on satellites.

Objects That Emit in the Infrared

Bound to the Earth or, at best, to satellites a few hundred miles above its surface, astronomers must discover the nature of celestial objects light-years away solely on the basis of the type and amount of radiation these objects emit. The picture cannot be complete without infrared observations. (Fig. 7.1) The amount of infrared radiation compared to the radiation emitted at other wavelengths is often critical for determining the physical nature of an object. In addition, many objects radiate the bulk of their energy in the infrared part of the spectrum, so only infrared measurements can tell how much energy is actually being emitted.

Infrared radiation is nothing more than light of a color redder than the human eye can detect.* In physical terms, this corresponds to infrared light having a longer wavelength. In fact, infrared light behaves very much like visible light; one uses mirrors and lenses to concentrate and focus it just as one would for visible light. Most astronomical infrared observations have been made with the same

---

* For details, see "Technical Note" at the end of this chapter.

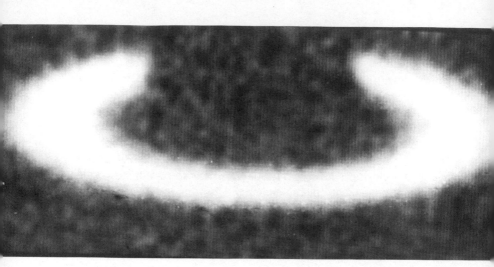

Figure 7.1: *An unfamiliar view of Saturn. At an infrared wavelength of 3.5 microns, only the rings are visible. The planet itself is a very poor reflector of sunlight at this wavelength because of absorbing gaseous methane in the planet's atmosphere. This picture, which illustrates how dramatically different our view can be at different wavelengths, was made by rapidly scanning the image of Saturn across a single detector and recording the signal. The total exposure time was 30 minutes with the 3.7-meter Anglo-Australian Telescope.* (Image by David Allen. Copyright Anglo-Australian Telescope Board. Used by permission.)

telescopes used at other times for visible observations, and even the few telescopes specifically designed for infrared observations differ from standard optical telescopes only in small (but important) details.

The longer wavelength of infrared light compared to visible light does lead to some important differences. One of the most important is the temperature of objects that emit the bulk of their radiation in each range of wavelengths. On Earth, infrared radiation is commonly associated with "warm" objects, but celestial objects that emit in this waveband are actually a great deal cooler than objects that emit visible light. This means that the kinds of objects that emit prominently in the infrared are not the familiar kinds of stars and galaxies that emit visible light. Indeed, much of the infrared emission we observe in the sky comes not from stars but from small, solid particles (dust) that are heated by some nearby object. Other sources of infrared radiation are associated with the most violent processes known, including extremely high-energy particles moving in magnetic fields and matter falling onto compact objects such as neutron stars or perhaps black holes.

Objects bright in the infrared typically have temperatures between 3 and 3,000 degrees Kelvin. Especially at the cooler temperatures, the emission is relatively small, and therefore to be detectable, an object must have an enormous surface area. This explains why dust is such a copious emitter of infrared radiation: finely divided material gives a very large surface area for the amount of mass present. In typical astrophysical environments, dust is heated to temperatures between 10 and 500 degrees Kelvin, just the right range to emit in the infrared.

One type of object that emits copious amounts of infrared radiation is a star near the end of its life (Figure 7.2). As the star approaches this stage, its outer layers expand and cool. For some reason that is poorly understood, material begins to flow from the star into the surrounding space at an enormous rate, often equivalent to blowing away the entire mass of our Sun in only 100,000 years. As gas flows away from the surface of the star at a rate 100 million times greater than the Sun's loss of mass through the solar wind, it eventually becomes cool enough for solid particles to condense. These particles, in turn, are heated by absorbing the visible light emitted by the star. This means visible light emerging from the dust shell is weakened, while the dust shell itself emits infrared radiation. The discovery that dust-enshrouded stars are relatively common forced an extensive revision of theories of stellar evolution. Although there was evidence for mass loss before infrared observations were made, it was sufficiently subtle to be ignored. Today, with this dramatic evidence before us, mass loss is recognized as a crucial part of stellar evolution, and all theoretical models must consider it.

Very young stars are examples of bright infrared sources at the other end of the stellar lifeline. As a giant interstellar cloud of gas and dust contracts under its own gravity, one or more stars begin to form. The details of the contraction process are still not understood, but shortly after the new star has formed, it is still surrounded by its natal cloud. The dust in this remaining cloud completely obscures visible light emitted by the newborn star. However, the dust itself emits strongly in the infrared, thus revealing the presence of a hidden star. Once the presence of the star is known,

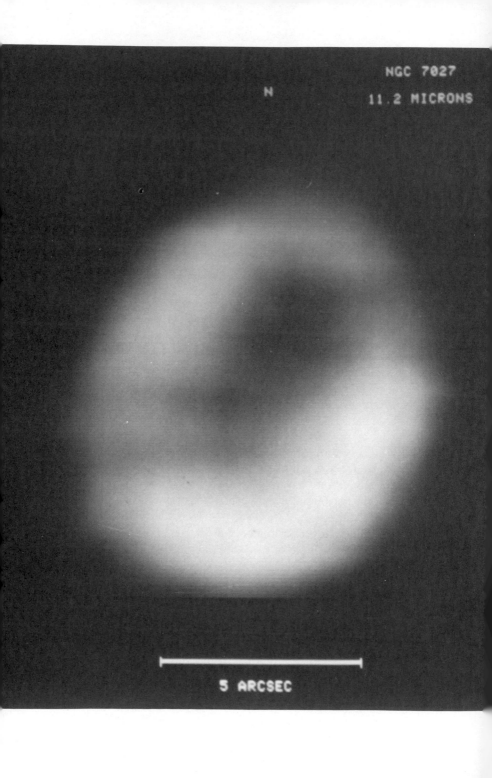

Figure 7.2: *Image of the planetary nebula NGC 7027 at 11.2 microns. Note the symmetric appearance typical of planetary nebulas, although the emission seen here is entirely from dust rather than the ionized gas that emits visible light. In visible light, this same object is partially obscured by dust, completely changing its appearance and leading to past suggestions that it might not be a planetary nebula! The central star, which would typically be bright in a visible-light photograph of an unobscured planetary nebula, is too faint to be seen in the infrared.* (Image obtained with the NASA-Goddard Space Flight Center's 10-micron array camera—a joint project of NASA-GSFC, Harvard-Smithsonian Center for Astrophysics, Steward Observatory of the University of Arizona, and NASA Ames Research Center—on the Infrared Telescope Facility, operated by the University of Hawaii.)

more detailed infrared studies can be undertaken to reveal the composition and motions of the star and the nearby material.

Galaxies are commonly described as large aggregations of stars, but they are much more than that. They include interstellar gas, dust, magnetic fields, high-energy particles, and at least one form of matter about which nothing is known except its existence. The nuclei of some galaxies ("active nuclei") show evidence of violent emission processes whose nature is poorly understood. One suggestion is that these nuclei may contain gigantic black holes. As matter falls into these gravitational wells, it is heated and emits radiation, including infrared radiation. If dust is present, it will absorb much of the shorter wavelength radiation and reemit it in the infrared. Indeed, some active nuclei have been discovered solely because of their infrared emission. (IRAS has probably found thousands more.) Infrared studies of active nuclei have revealed that many are much more luminous than was deduced from optical observations, simply because the bulk of their radiation is in the infrared. Also, in many nuclei, the dust substantially weakens and modifies the emerging visible radiation, so corrections must be made in deducing properties of the nuclei.

Other Advantages of Infrared Observations

The infrared region of the electromagnetic spectrum has two other characteristics particularly important for astronomy. First is the ability of infrared light to penetrate dense interstellar clouds of dust; second is the fact that many key atomic and molecular constituents of the cosmos emit and absorb radiation primarily at infrared wavelengths.

The superior penetrating capability of infrared light is a direct result of its longer wavelength compared to visible light. Dust particles cannot interact efficiently with radiation having a wavelength much larger than the particle circumference. Since most interstellar particles have diameters in the range of 0.1 to 0.5 micron, the visible wavelengths of 0.4 to 0.8 micron are absorbed, while the longer infrared wavelengths are transmitted. Dust between the Earth and the center of the Milky Way galaxy, for example, allows only one ten-billionth of the visible light to reach our telescopes. This tiny fraction is completely unobservable. On the other hand, at a relatively short infrared wavelength of 2 microns, about one-tenth of the light reaches us. This amount is quite sufficient to make the galactic center one of the brightest infrared sources in the sky. Similarly, the dusty interiors of the regions where stars are actually forming remain well-hidden in visible light, but can be seen in the infrared. There are many other examples of interesting objects hidden behind dust that visible light cannot penetrate and that causes a large diminution in the brightness of the source. Infrared observations are essential for studying such objects (Figures 7.3 and 7.4).

All atoms and molecules emit and absorb radiation at particular wavelengths. These wavelengths can range from the radio to the X-ray or even the gamma-ray part of the electromagnetic spectrum. Thus, to study a certain atom or molecule, one must observe the distinctive wavelength where it emits or absorbs. As laboratory chemists know well, the infrared region is particularly important for many molecules. Astronomers, too, know that such cosmically abundant and important molecules as H_2, CO, H_2O, CN, and many others can be best studied at infrared wavelengths. Even when the molecules can also be studied at radio wavelengths (and the most abundant molecule, H_2, cannot!), infrared observations usually provide different and complementary information about the location, motion, and physical state of the molecules. In addition, individual atoms and ions (atoms with one or more electrons removed) also can be studied in the infrared. Many of these atoms and ions cannot be studied at all at visible wavelengths, or the

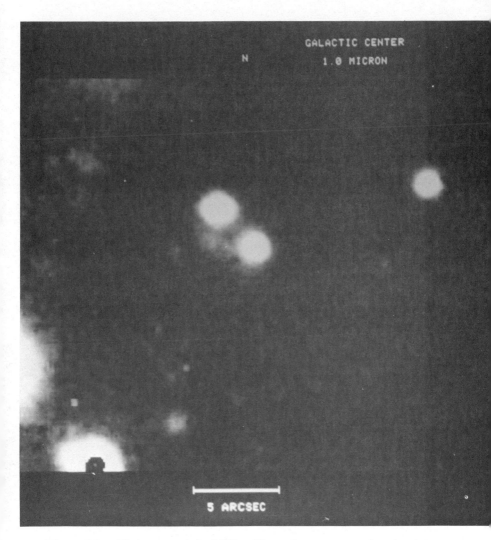

5 ARCSEC

Figure 7.3: *The center of the Milky Way galaxy at a wavelength of 1 micron, as seen by a visible-light camera used at its extreme red limit. The two bright objects near the center are foreground stars, but the faint wisps around them are probably star clusters near the galactic core.* (Courtesy of J. P. Henry, D. L. DePoy, and E. E. Becklin)

observations may require large and uncertain corrections.

Finally, the interstellar dust grains themselves often produce characteristic signatures of their composition at infrared wave-

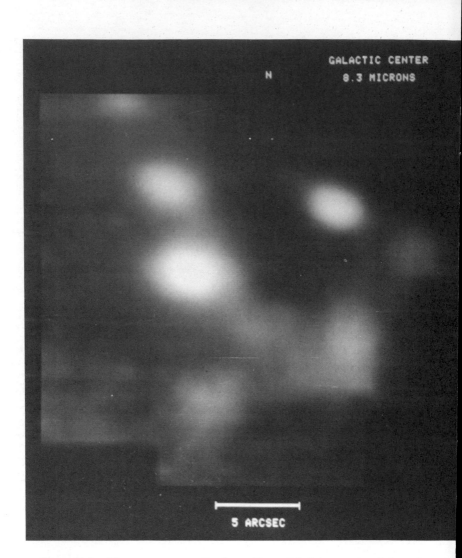

5 ARCSEC

Figure 7.4: *The same area on the sky as Figure 7.3, but now at an infrared*
wavelength of 8.3 microns. It is a composite of smaller images made with
an array camera. The bright objects are dust clouds near the galactic center.
(Image obtained with the Goddard Space Flight Center's 10-micron array
camera)

lengths. Earth rocks, for example, are overwhelmingly composed
of silicates, which emit and absorb radiation with particular efficiency at wavelengths near 10 microns; and, indeed, observations

near this wavelength show that interstellar dust also contains a substantial fraction of silicates. Other dust constituents include very small carbon particles, water ice, and probably complex organic (i.e., carbon-containing) molecules.

How to Detect Infrared Radiation

Most practical detectors for infrared light fall into one of two groups: *quantum detectors* or *thermal detectors*. Each type makes use of a different property of electromagnetic radiation to convert infrared light to an electrical signal. In each case, the electrical signal is proportional to the intensity of the light falling on the detector.

Thermal detectors are the simplest type. They are really just thermometers. When energy of any sort is absorbed by the detector, it heats up. The change in temperature changes some other property of the detector, usually its electrical resistance, and that change is recorded. A thermal detector of this type is called a *bolometer*. There are two requirements for measuring the tiny amounts of radiation reaching the detector from distant celestial objects. First, the detector must have a very small heat capacity so that a given amount of energy produces a large change in temperature. Second, the detector must be operated at a low temperature so that any small absolute change in temperature represents a relatively large fractional change and thus produces a significant change in electrical resistance. These requirements are met by using a very small semiconductor element as the bolometer and cooling it to the lowest practical temperature. Generally, a liquid-helium container is built into the instrument to reach temperatures as low as 1 degree Kelvin. (A more complicated scheme involves using the light isotope of helium [^{3}He], and temperatures below 0.4 degree Kelvin can then be achieved.)

Bolometers have the advantage of being sensitive to every kind of radiation that they can absorb. This means they are useful in devices that must detect radiation over a wide range of wavelengths or at wavelengths where other types of detectors are not available. At present, bolometers are generally used for wavelengths between 10 and 1,000 microns (= 1 millimeter). In theory, bolometers are the best radiation detectors, but in reality, their sensitivity is limited by the temperature to which they can be cooled. For many applications, other types of detectors are superior.

Quantum detectors are the type most used by astronomers. In such a device, a single photon striking some material causes a measurable event. Successive events are simply counted, and their rate is proportional to the brightness of the light falling on the detector. A photon of visible light can actually eject an electron from a metal in a process called the *photoelectric effect*. The ejected electron can be amplified and detected essentially without noise. In the infrared, photons have energies too low to eject electrons entirely. Instead, electrons are "excited," that is, boosted to a higher energy state inside a semiconductor crystal. Electrons in this higher state carry an electric current, so the current in the crystal is proportional to the brightness of the light falling on the crystal. In most astronomical observations, photons are few and the resultant small currents must be highly amplified to be measured. When amplification is done by standard electronic techniques, noise is inevitably introduced in the measurement. (In addition, most materials allow some current to flow even when no light falls on the detector. This current is appropriately called *dark current*, and it, too, introduces noise.)

Individual detectors can measure the brightness of only a single region of the sky at one time. Typically, an astronomer chooses a region that includes an interesting object and then measures the total brightness at one wavelength. If the object is not a distinct point source, however, building up a map or image can be a tedious process. For a large, diffuse object such as a dust cloud, each point on the object must be measured separately and the individual measurements calibrated and combined to form an image. In the past, images were built up by slowly scanning the whole telescope across the object. More recently, techniques for rapid scanning have been developed. (One result is shown in Figure 7.1.) Still, no matter how rapidly one scans, a single detector can study only one part of an object at a time, and the total observing time available must be divided among all the parts. With the exposure time on any one position so limited, only relatively bright objects can be mapped.

An immense improvement in infrared detector technology has very recently become available. By applying advances in semiconductor technology to photodetectors, it has become possible to combine many detectors into an array. Astronomers have for years used similar detector arrays to make measurements in visible light, and the same technique can now be extended to the infrared.

The great advantage of detector arrays is that each element is looking at its part of the object all of the time. The time necessary to build up an image is thus decreased by a factor equal to the number of detectors in the array. If an array is 60 elements square, an image (or a spectrum) that would normally take an hour to produce with a single detector can be made in one second. By implication, this means much longer exposure times, and thus studies of much fainter objects are now practical. (Even better, certain types of detector arrays actually have less noise per element than single detectors of the same type.) An additional advantage is that the relative positions and brightnesses of different objects in one image are easier to determine. Some examples of images made with detector arrays are shown in Figures 7.2, 7.3, and 7.4. Although not yet in common use, infrared detector arrays could produce advances comparable to the introduction of photography into optical astronomy.

Telescopes for the Infrared

Obviously, it is easier to measure a faint source if the light from any other sources can be eliminated. After all, it is much easier to see the stars at night than in the daytime. Infrared astronomers face a special challenge in eliminating extraneous light because the atmosphere, the inside of the dome, and even the telescope itself are all emitting infrared light simply because they are at the right temperature. Light from the inside of the dome can be excluded by enclosing the detector in a cooled chamber that admits only the light coming through the telescope, but infrared radiation from the atmosphere and the telescope itself are not so easily eliminated.

The infrared radiation of the atmosphere and telescope can be described by a quantity called the *emissivity*. This is simply the ratio of the actual emission to the theoretical emission of a "perfect radiator" at the temperature of the telescope. The emissivity is easily measured when an infrared detector is mounted on a telescope simply by comparing the detector output with the dome open to the output with the dome closed. Having the dome closed is equivalent to a perfect radiator, while the telescope mirrors and atmosphere are much less efficient radiators.

Although telescopes can be designed to have low emissivity, very few large telescopes have been. Most were built before infrared detectors reached their current sensitivity, and neither the need for low emissivity nor the means of achieving it were widely recognized. Also, a general-purpose telescope used for both optical and infrared astronomy must meet other design goals. For example, good baffling of stray visible light and a wide field of view may benefit the optical astronomer, but they will also increase the emissivity.

Actually, the design of a low-emissivity telescope is not complicated in principle. Since highly polished surfaces emit much less efficiently than dull surfaces, one simply arranges for the detector to receive as little light as possible from anything except the telescope mirrors. Telescope mirrors may also be coated with silver or gold rather than the traditional aluminum, since these metals have lower emissivity. (Fortunately, mirror coatings are very thin, and even taking into account the inefficiency of the coating process, less than 100 grams of gold would be needed to coat the primary of the 4-meter reflector at Kitt Peak. This may sound like a lot, but the cost of the material is still much less than the labor cost.) In principle, one could design a telescope with emissivity less than 5 percent. In practice, however, dust on the telescope mirrors emits much more than the mirrors themselves. A good example is the 3-meter infrared telescope facility located on Mauna Kea, Hawaii, which was designed for low emissivity and is used almost exclusively for infrared observations. Its emissivity usually varies between about 10 and 25 percent, depending on how recently the mirrors have been washed. Typically, general-purpose telescopes may have emissivities between 20 and 40 percent.

The emissivity of the atmosphere depends on its transparency, with the emissivity increasing as transparency decreases. Since water vapor is a major absorber, it is doubly important to locate an infrared telescope at a dry site. High mountains are best because water vapor is strongly concentrated toward low altitudes. Typical atmospheric emissivities at wavelengths routinely observed from the ground are about 5 percent.

New Ground-based Telescopes

For many newly built telescopes, especially the larger ones, infrared observations form a major part of the research program, and the designers strive for low emissivity. One good example is a new 4-meter-diameter telescope proposed for the Cerro Tololo Interamerican Observatory (CTIO), located near La Serena, Chile, and funded by the U.S. National Science Foundation. The existing 4-meter telescope at CTIO is well adapted to wide-field photography and visible spectroscopy, so the new instrument will be optimized for infrared observations.

Because the instrument would be designed for specific observations rather than as a general-purpose instrument, the cost would be relatively low. Several technical innovations would also keep the construction costs down. Rather than a conventional equatorial mount, the telescope would use an alt-az mount, which reduces the size and thus the cost of the mounting. Another cost saving comes from use of a primary mirror with a short focal length, which reduces the size and cost of both the mount and the dome. The short focal length also reduces the area on the sky that can be observed at one time, but this is no problem for the kinds of observations proposed because the area on the sky will be limited by the detectors, not the telescope.

If it is built, the CTIO telescope will be the largest infrared-optimized telescope in the Southern Hemisphere, capable of observing many important southern objects, including the center of the Milky Way, the Magellanic Clouds, and a host of other objects not seen by northern telescopes. The estimated cost is $10 million, and the completion date is estimated as 1988. For comparison, it would probably cost $30 million to $50 million to duplicate the existing 4-meter telescope. Even if this particular telescope is not built, similar ones certainly will be.

A much more ambitious project is the National New Technology Telescope (NNTT) (see page 39). Infrared performance has been one of the most important criteria in the design of this telescope, and the current specification requires that the NNTT have an emissivity of less than 5 percent. Obviously, the NNTT will be located at a very dry site on top of a tall mountain. And the 10-meter Keck Telescope of Caltech and the University of California, planned for

Mauna Kea, will also be used for infrared astronomy (see pages 36–39).

The need for large telescopes is even more critical in the intrared than in the visible part of the spectrum simply because the size of the telescope limits the size of the finest detail that can be studied.* This limit is called the *diffraction limit*, but for visible light, it is usually less important than the limit set by turbulent air currents in the atmosphere above the telescope. Astronomers call this limit *seeing*. At good sites on the best nights, the optical seeing can be better than 0.5 second of arc, but a limit of 1 arcsecond is more common. The seeing is better in the infrared because the longer wavelengths are less affected by small-scale atmospheric turbulence. The diffraction limit for a telescope of a particular size becomes worse, however, and at an infrared wavelength of 10 microns, even a 4-meter telescope can produce images no smaller than 1.2 arcseconds in diameter. The only way to study finer details is to use a larger telescope. The diffraction limit for the NNTT at 10 microns will be 0.3 arcsecond, which should allow study of the smallest details permitted by the atmosphere.

Airborne Telescopes

Because the Earth's atmosphere blocks radiation from celestial objects over a wide range of wavelengths, there have been numerous efforts to lift telescopes to as high an altitude as possible. Aircraft, balloons, and rockets have all been used. Of these, only rockets get above the atmosphere entirely, but observing time on them is limited to a few minutes per flight. Balloons can rise above nearly all of the atmosphere, although enough remains to screen out a few important wavelengths. Research groups in many different countries use balloons for infrared astronomy, and telescopes with mirrors up to 1.2 meters in diameter have been flown successfully. A typical research group may conduct from one to a few flights per year, with each flight lasting 8 to 10 hours.

At present, the most frequently used high-altitude infrared telescope is the one aboard the Kuiper Airborne Observatory (KAO).

---

* There is a very simple formula that relates the dimension of the telescope to the angular size of the blurred image of a negligibly small source. The blur circle diameter in radians is just 2.4 times the wavelength divided by the diameter of the telescope.

Figure 7.5: *The Kuiper Airborne Observatory in flight. The telescope is located inside the cavity just forward of the wing. The telescope is movable in elevation angle, and the direction of flight controls the azimuth angle. Both angles must be adjusted continuously in order to track a particular celestial object.* (Photo courtesy of NASA/Ames Research Center)

NASA built the KAO about 1974 by cutting a hole in the side of a C-141 military transport airplane and putting a 0.9-meter (36-inch) telescope inside (Figure 7.5). The telescope is isolated from the aircraft body by an air-flotation bearing and controlled by gyros and star trackers to point with a steadiness comparable to that of a ground-based telescope. It is pointed by being rotated about a horizontal axis to set the elevation, while the entire airplane turns to set the azimuth. About six hours of research can be accomplished on each flight, and since the first flights in 1974, NASA has been supporting about 70 research flights per year. (This number could be increased to 120 if the budget permitted.)

Observing on the KAO is strikingly similar to observing at a ground-based telescope. An observing team arrives a few days before the scheduled flight, installs an instrument that it has built, tests everything possible, and establishes a schedule of the objects to be observed. The instrument is generally installed in the cabin of the aircraft so the team can make necessary adjustments—or give a balky instrument an appropriate thump. During the flight, the team uses its instrumentation and the computers provided by the KAO to acquire and analyze data. If a team has more than one flight scheduled, the plans for subsequent ones can be altered depending on the data already acquired.

Observing on the KAO does have some disadvantages. The aircraft cabin is cold and so noisy that headsets must be worn continuously. The in-flight meals are typical of airline food, so the experienced observer usually stops off at a local deli before the flight. More important scientifically, the decision of what objects to observe and the amount of time to spend on each must be made well before the flight—with little chance for last-minute changes to follow up unexpected discoveries. Moreover, since the direction of flight depends on the object being observed, the observing list must include objects in different directions so that the aircraft will return to the base before it runs out of fuel. (Thankfully, this requirement is strictly enforced!)

The achievements of KAO observers are numerous. They include discovery of the rings around Uranus, analysis of the composition of the atmosphere of several planets, discovery of new constituents of interstellar dust and gas, and measurement of the infrared luminosity of many types of objects. Over the years, the observations conducted have changed from the "new discovery" type to the "detailed analysis" type. The key factor in this change has been the ability to plan and make high-altitude observations with almost the same ease as those made from the ground.

Both balloons and rockets reach much higher altitudes than aircraft, but the astronomer does not go along. This makes it much more difficult to utilize complex instruments. All operations must be by remote control, and for rockets, the entire sequence of observations must be planned in advance. Nevertheless, the lower temperature, reduced atmospheric cover, and the improved sensitivity have attracted hardy astronomers to rockets and balloons. Indeed, before IRAS was launched, rockets had performed surveys of the brightest objects over nearly the entire sky. The most striking results from balloons have been maps of giant star-formation regions and other areas of the sky larger than can be studied with the KAO.

Airborne telescopes larger than those now in use could offer important and unique observational capabilities. A preliminary design study has shown that it would be possible to build a 3-meter telescope in a Boeing 747-SP aircraft, using techniques demonstrated by the KAO. A design study for a 3-meter balloon-borne telescope also is now in progress. It is hard to predict whether either of these telescopes will ever be built. A 3-meter telescope in an airplane would probably cost between $50 million and $100 million, depending on the cost of the airplane, but it could provide perhaps 1,000 hours of observing per year. A 3-meter balloonborne telescope would be substantially cheaper, but it would provide much less observing time.

Telescopes in Space

Even under the best conditions, ground-based telescopes will probably always have emissivities of 10 percent or more. Doing infrared observations under these conditions is very much like trying to do visible-light observations in twilight: bright sources can be seen but not faint ones. Our knowledge of the universe would be extremely limited if optical astronomers were restricted to observing only at twilight, yet such is the situation in the infrared today.

Essentially complete elimination of telescope emission is possible if the telescope is cooled. To take full advantage of this, however, it is necessary to put the telescope into space. Aside from practical problems such as frost, a cooled telescope on the ground would still suffer from atmospheric emission. In space, however, the only interference is radiation from the warm—but very sparse—interplanetary dust. This corresponds to an emissivity of about 0.00001 percent, or at least a million times better than for a ground-based telescope (Figure 7.6). Naturally, all atmospheric absorption would be completely eliminated also.

The first satellite to take advantage of the vast sensitivity increase possible with a cooled telescope was the infrared astronomical satellite (IRAS). It was launched in January of 1983 and ran out of liquid-helium coolant in November of that year. During its 10 months of operation, it surveyed almost the entire sky at wavelengths between 12 and 100 microns. Even though it carried only a 60-centimeter (24-inch) telescope, IRAS was so sensitive that it measured over a quarter of a million objects. By comparison, fewer than 5,000 had been measured at these wavelengths in the entire previous history of astronomy. Moreover, most of the previously

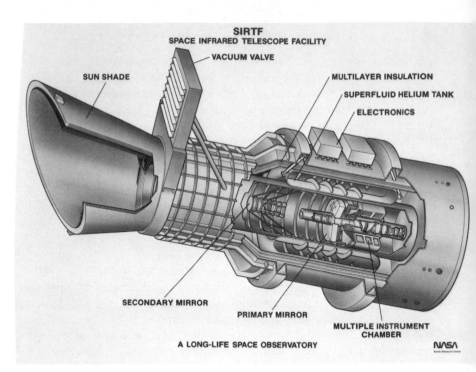

SIRTF
SPACE INFRARED TELESCOPE FACILITY

VACUUM VALVE

SUN SHADE

MULTILAYER INSULATION

SUPERFLUID HELIUM TANK

ELECTRONICS

SECONDARY MIRROR

PRIMARY MIRROR

MULTIPLE INSTRUMENT
CHAMBER

A LONG-LIFE SPACE OBSERVATORY

NASA
Ames Research Center

Figure 7.6: *Comparison of the sensitivity of astronomical instruments expected for the 1990s. The horizontal axis represents wavelength, with longer wavelengths to the left; the vertical axis represents source brightness. The solid lines indicate the wavelength coverage of the various instruments as well as how faint a source each can detect. The upper dashed line shows the brightness (at all wavelengths) of the relatively nearby quasar 3C273. The lower line shows an estimate of the brightness of a similar quasar located at a much greater distance. The instruments depicted are: the Very Large Array (VLA); the Infrared Astronomical Satellite (IRAS); the Space Infrared Telescope Facility (SIRTF); the Hubble Space Telescope (ST); the Advanced X-ray Astrophysics Facility (AXAF); and the Gamma-Ray Observatory (GRO). (The Infrared Space Observatory [ISO] will have wavelength coverage and sensitivity similar to SIRTF.) Of all proposed infrared instruments, only SIRTF and ISO are sensitive enough to study the faint objects that will be observed at other wavelengths.* (Diagram courtesy of NASA/Ames Research Center)

measured objects were known from studies at other wavelengths. Thus, IRAS detected for the first time many new types of objects that are especially bright in the infrared.

Another space mission that is similar but complementary to IRAS is the small helium-cooled infrared telescope (IRT) on Spacelab 2, a NASA Space Shuttle mission in 1985. The wavelength range studied, roughly 7 to 95 microns, will extend the IRAS survey to slightly shorter wavelengths. Specifically, the IRT was designed to study large-scale structures in the infrared sky caused by dust in interplanetary space, within the Milky Way, and outside of the galaxy. The IRT also conducted engineering tests related to the future design of space infrared telescopes, including measurement of possible contamination of the observing environment caused by the Space Shuttle itself.

After IRT, the next infrared satellite will be the Cosmic Background Explorer (COBE), which is equipped with three cooled telescopes operating over the wavelength range from 1 micron to 1.2 centimeters. COBE will measure all diffuse sources of radiation within its wavelength limits. Although interplanetary and interstellar dust will be measured, the main interest is to study extragalactic and cosmological sources. One such source is believed to be a remnant of the Big Bang—the primeval fireball that astronomers think began the expanding universe. COBE will measure the uniformity and exact wavelength dependence of this radiation, characteristics crucial to understanding its origin and determining the conditions prevailing in the infant universe. Galaxies in their earliest stage of star formation might also be detected, thereby setting the temperatures for the first generation of stars and the dust content of primitive galaxies. COBE will be carried into orbit by the Space Shuttle and then boosted into a higher orbit by an attached rocket. Launch is now scheduled for 1987.

Although it was hugely successful, the IRAS mission had several limitations that will not be overcome by either the IRT or COBE. In fact, the main limitation of all three missions is that they were designed to conduct surveys rather than to carry out detailed study of individual sources. (Of course, it made good sense to find out what kinds of objects exist before proposing to study them in detail!) Another limitation is that both IRAS and IRT operated at only four particular wavelengths; for more detailed studies, the complete spectra of infrared sources are necessary. Yet another limitation is spatial resolution. The IRAS telescope measured the total amount of light in squares 40 arcseconds on a side; the IRT will measure areas roughly 0.6 degree square. For comparison, visible, radio, and ground-based infrared observations can study the structure of sources on a scale of 1 arcsecond. Still another limitation is the observing time on each source. The IRAS telescope scanned over each source in less than a quarter-second. Just as long-exposure photographs detect fainter objects than short exposures, increased infrared observing time results in greater sensitivity. (IRAS did make some long-exposure observations, but only for relatively few sources.) Finally, the IRAS mission was limited in time—just 10 months with no repetition. As astronomy advances, there will be a continuing need to observe new types of objects and to continue to monitor dynamically changing objects with the sensitivity that only cooled telescopes can achieve.

There are currently two projects that offer the potential of long-term and detailed observations with cooled telescopes in space. These are Infrared Space Observatory (ISO), a project of the European Space Agency, and Space Infrared Telescope Facility (SIRTF), a NASA project (Figure 7.6). The projects are similar in some respects, but there are several important differences. The ISO mission is conceived as a single flight of approximately 18 months' duration. The satellite would be placed in a 12-hour orbit to allow relatively long uninterrupted observations, but such an orbit is too high for the satellite to be recoverable by the Space Shuttle. The SIRTF, on the other hand, is envisioned as a more or less permanent observatory. It must therefore be in an orbit that the Shuttle can reach (probably with the assistance of a new "orbital maneuvering vehicle") to allow periodic replenishing of the coolant supply and the replacement of components that will wear out. An attractive possibility would be to perform this "refueling" operation at the proposed Space Station. More complicated maintenance tasks also could be performed. This replacement of coolant and reuse of the spacecraft could be repeated as long as the telescope continues to serve a need.

The design and instrument complements for ISO and SIRTF also reflect their different missions. The SIRTF telescope diameter will be 0.85 meter versus 0.6 for ISO. SIRTF's pointing accuracy will be better, too, and it will make more use of two-dimensional detector arrays. In short, SIRTF will have better ability to survey

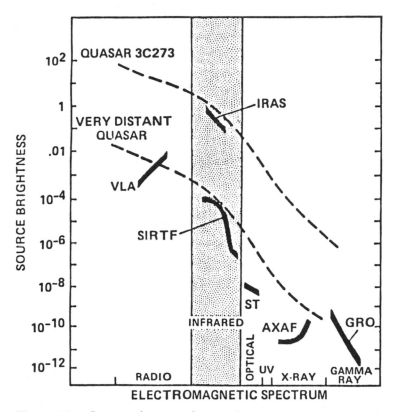

Figure 7.7: *Cutaway diagram showing the proposed design for SIRTF. The entire telescope volume inside the "multilayer insulation" will be cooled to the temperature of liquid helium to eliminate thermal radiation from the telescope. The "multiple-instrument chamber" will contain cameras and spectrometers, to be built separately from the telescope and installed before launch.* (Artist's conception, courtesy of NASA/Ames Research Center)

regions for previously unknown sources, to study fine spatial detail, and to study fainter sources. On the other hand, the ISO instruments are better adapted for detailed study of the spectra of individual sources.

Predictions about the schedule and cost of space projects are difficult, since both depend more on political factors than scientific or technical ones. ISO is an approved project, and studies are now in progress to produce a final description of telescope and instrument capabilities. Equivalent studies for SIRTF are not yet funded, and the mission cannot receive final approval until those studies are complete. At this writing (May 1985), NASA expects to start the SIRTF instrument studies in October 1985 and the SIRTF telescope study in early 1986. Whether or not that is possible depends on what budgets pass Congress. Exact costs of ISO and SIRTF cannot be estimated until the studies are complete, but each mission should be in the range of hundreds of millions of dollars.

In the deserved enthusiasm for cooled telescopes, one should not neglect the infrared possibilities of the Hubble Space Telescope (HST). The first instrument complement will not include any infrared capability, but this lack is mostly a result of the relatively backward state of infrared technology in 1977 when the instruments were selected. However, the importance of infrared observations was recognized, and the design of HST provides for the later incorporation of an infrared instrument.

HST actually has important advantages over other potential infrared instruments. The major one is its capability for studying fine details in images, thanks to its relatively large mirror placed above the Earth's atmosphere. Another important advantage is the ability to observe at any wavelength, not just at those wavelengths where the Earth's atmosphere is transparent. This is especially important for the study of many atoms, ions, and molecules that emit and absorb radiation at fixed wavelengths unobservable from the ground. One limitation of HST is that the mirror is not cooled; in fact, the mirror is heated to approximately room temperature in order to maintain its precise shape. Another limitation is the 2.3-meter aperture; several ground-based telescopes 3 to 6 meters in diameter exist now, and larger ones are planned. However, the HST was designed for low emissivity, there is no atmospheric emission, and there is very little dust to dirty the mirror. On balance, the HST's infrared sensitivity should be about the same as for ground-based telescopes, and it will be the best instrument for achieving high spatial resolution with unlimited wavelength coverage.

In the more distant future, there is the possibility of building very large telescopes in space. The concept likely to be built first is known as the large deployable reflector (LDR), a 20-meter-diameter dish shipped into space in parts by one or more Space Shuttles. Once in orbit, astronauts would remove LDR from the Shuttle's cargo bay, reassemble it into its original, highly precise shape, and leave it in orbit. LDR will not be cooled, so it will not be able to observe objects as faint as can ISO or SIRTF, but it will give considerably sharper images. Many technical problems must be solved before LDR is a reality, but it is envisioned as a project for the late 1990s.

Theoretically, at least, it should be much easier to build large telescopes in space than on the Earth. In the reduced gravity, the required rigidity and stability can be maintained over much larger dimensions. The many practical problems, however, probably will prevent large telescopes from being constructed in space until it is routine to build many other kinds of structures there. When this will happen is anyone's guess, but I will take a chance and predict that, by the year 2005, it will be cheaper to build a 15-meter telescope in space than on the ground. On the other hand, the technique of *interferometry* is advancing so rapidly that, by 2005, large single-mirror telescopes may not be as useful as interferometers of comparable cost. Indeed, interferometers with dimensions of tens, hundreds, or even thousands of meters may then be built (see Wesley Traub's chapter). Regardless of what instruments are used, we can be sure that infrared observations will be even more important for understanding astrophysical objects.

What We May Learn

In the future, infrared observation will contribute to all areas of astronomical research. As improving technology makes fainter objects observable, not only will many more objects of a given class be measured, but also whole new classes of objects will come within the measurable range of brightness. Many quantities now measured relatively poorly with visible light will be measurable much more accurately in the infrared. For example, already one of the best distance indicators to external galaxies is based on infrared observations, and improved instruments should allow measurement of many more galaxies. Studies of Solar System objects, individual stars, star clusters, planetary nebulas, and supernovas will all benefit.

The most exciting prospects, however, are the wholly new kinds of observations that will be possible. For example, we may detect galaxies still in the process of formation. By looking at very distant objects, we are also looking back in time. If we can see far enough, we should be able to look back on the era when galaxies first began to contract under their own gravity to form distinct structures. No one really knows what they will look like, but they are most likely to be found in the infrared.

Another possibility is the direct detection of objects smaller than the stars now known. The distinction between a "large planet" and a "small star" may not be a clear one, but there is currently no way of knowing if such objects exist. One technique for finding low-mass objects is to look for small companions of the nearest stars. Objects slightly larger than Jupiter should be detectable by SIRTF, and dust shells similar to the interplanetary dust in our own system may also be seen. Astronomers might even search for low-mass objects collectively rather than individually by carefully observing nearby galaxies. Galaxies are known to contain mass in some form invisible to existing instruments. If giant planets or very-low-mass stars (brown dwarfs) constitute a substantial fraction of this hidden mass, SIRTF should be able to detect them.

The history and occurrence of star formation in different types of galaxies can be explored by infrared observations. Star-formation regions glow brightly in the infrared because the embedded stars heat the surrounding dust, and the brightest regions of a galaxy will be those in which stars are currently forming. Because optical pictures can locate existing stars and radio pictures can locate gas clouds, comparison with infrared images should reveal the conditions necessary for the birth of stars. Past star formation can also be traced by the distribution of heavy elements in galaxies, since these elements are formed in the interiors of stars and are then ejected. Visible light measurements are both indirect and limited in accuracy, but infrared measurements can yield the heavy element abundance more directly. New telescopes and instrumentation will provide the required sensitivity.

In any immature science like astronomy, the most important discoveries are usually those we do not foresee. Whenever a new range of wavelengths is explored, altogether new phenomena are likely to be discovered. In view of the very preliminary state of our current knowledge about the infrared sky, there is every reason to expect many more surprises.

Technical Note

Every macroscopic object in the universe emits radiation, provided its temperature is above absolute zero. (The situation for individual atoms and molecules is more complicated because the definition of temperature may not be straightforward. Any object with more than a few hundred atoms can be considered macroscopic for our purposes.) In general, the type of radiation emitted depends on the size of the object, its composition, and its temperature. Size is important only in that the object must be larger than the wavelength of interest. Composition can be very important, but many objects of interest in astrophysics are mixtures of many materials, and the effects of differing composition are minor compared to temperature. Thus, temperature is often the determining influence on what type of radiation a body will emit. Because of this, any radiation a body emits merely because it is above absolute zero is called *thermal radiation*. This radiation can emerge at any wavelength, depending on the temperature of the emitting object, but for objects near room temperature, the thermal radiation emerges primarily in the infrared. The graphs in Figure 7.8 show the emission at various wavelengths for objects of common temperatures.

For an object to be observed by virtue of its thermal emission, it must be above some minimum temperature. This is a familiar phenomenon in the visible: room temperature objects (e.g., cats, chairs) do not appear to glow, but hot objects (e.g., light bulb filaments, the Sun) do. However, cooler objects are still sufficiently hot to emit radiation at longer (i.e., infrared) wavelengths, and if our eyes were sensitive to wavelengths 10 times longer, we would see all ordinary room temperature objects glowing. Certain snakes (the pit vipers) do in fact have sense organs for infrared radiation, which they use to find their warm-blooded prey through its higher thermal emission.

The laws governing the temperature dependence of thermal radiation are actually rather simple. They can be summarized as follows:

1. Thermal radiation emitted by a body is proportional to its surface area.
2. At a given temperature, the amount of radiation emitted increases

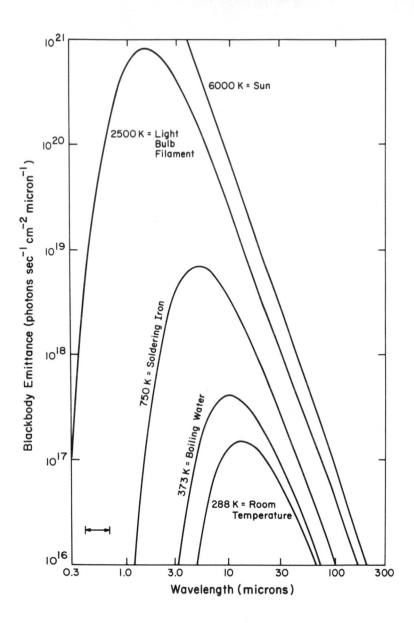

Figure 7.8: Thermal radiation from an object that emits perfectly. Wavelength is represented along the horizontal axis. The height of the curve is proportional to the rate at which photons are emitted at each wavelength. The curves are designated by temperature measured in degrees Kelvin (K). The two lines with arrows between them show the approximate wavelength limits of visible light.

toward shorter wavelengths until a maximum is reached. At still shorter wavelengths, emission falls very rapidly.
3. Hotter objects have their maximum emission at shorter wavelengths. This is known as Wien's Law, and is familiar to anyone who has watched objects being heated to high temperatures. An electric-stove burner, for example, changes from black to dull red to orange as it heats up. The filament of an incandescent light bulb is even hotter and appears almost white. In short, as the temperature increases, progressively shorter wavelength radiation is emitted, and our eyes perceive the light to be less and less red.
4. The amount of thermal radiation emitted by a body at any wavelength increases as the temperature increases. At wavelengths longer than the maximum emission, the increase is relatively slow. At wavelengths just shorter than the maximum, however, a small increase in temperature leads to a very large increase in emission. This explains how the bright lights used for making television shows and movies work: their filaments are only slightly hotter than those in normal incandescent light bulbs, but visible wavelengths are shorter than the peak emission, and the small temperature change results in a large change in brightness.

The complete formula for the temperature and wavelength dependence of thermal radiation was first given by the great German physicist Max Planck early in this century. The desired result, designated $N\lambda$, is the number of photons emitted each second per unit surface area per unit wavelength interval. In the formula below, λ denotes the wavelength, T the temperature, c the speed of light, π and e the usual mathematical constants, and h and k the constants of nature known, respectively, as Planck's constant and Boltzmann's constant. In fact, Planck introduced his now famous constant in this very formula:

$$N\lambda = 2\pi c\lambda^{-4} [e^{hc/\lambda kT} - 1]^{-1}.$$

It is easy to see how the curves in Figure 7.8 were calculated. At long wavelengths, the term in brackets reduces to $hc/\lambda kT$, and the total emission is proportional to λ^{-3}. We thus see a rapid increase toward short wavelengths. At short enough wavelengths, however, the exponential term in the denominator gets large, and the emission decreases rapidly.

The wavelength of maximum emission occurs at a particular

value of the product of wavelength and temperature and thus is inversely proportional to temperature. (This explains Law 2 above.)

The total power radiated at all wavelengths can be shown to be proportional to the fourth power of temperature. This is known as the Stefan-Boltzmann Law, and the constant of proportionality as Stefan's constant, or the Stefan-Boltzmann constant.

Probing the Sun's Secrets

An Advanced Solar Observatory

GEORGE L. WITHBROE

The Sun is an undistinguished, somewhat pedestrian, middle-aged star no different than countless billions of others in the universe. However, since it is also the star that lights up our days and makes life on Earth possible, it is rather important to us.

The Sun is also significant as an astronomical object, since it is the only star close enough to allow observations of its surface details. Because it is so near to Earth, we can probe it with a wide range of techniques using instruments sensitive to optical, radio, X-ray, and gamma-ray emissions. Since terrestrial weather as well as the conditions in the space immediately surrounding our planet are controlled by the amount and form of the energy received from the Sun, it is important to determine, and ultimately predict, how the Sun's energy production varies over both short (minutes to days) and long (months to centuries) time scales. Moreover, by probing the Sun we can also learn much about the fundamental physical mechanisms that drive and produce variations in the vast energy output of all stars, energy that escapes as radiation, out-flowing gas, and high-energy particles (cosmic rays). To accomplish all this, we must understand the physics of the Sun—how this relatively dull, yet still remarkable, star works.

Despite its nearness to Earth and its accessibility to ground-based telescopes, many of the most interesting phenomena occurring on the Sun require observations from experiments flown on rockets and orbiting spacecraft. There are two basic reasons for this. First, the Earth's atmosphere is opaque to critical sections of the solar spectrum, such as the far ultraviolet and X-ray regions. Second, motions in the Earth's atmosphere wash out the fine details of the solar surface. (These are the same motions, called *seeing* by astronomers, which cause stars to twinkle.) Because of this, we can neither observe fine structural details on the solar surface nor observe critical aspects of those phenomena, such as solar flares, which produce most of their energy at wavelengths absorbed by the terrestrial atmosphere. For the future, then, many solar problems will be studied—and solved—only by experiments on orbiting spacecraft.

Magnetic Nature of the Sun

The Sun is a huge ball of gas approximately 860,000 miles in diameter, or some 100 times the diameter of the Earth. It is about 4.6 billion years old, approximately halfway through its life as an ordinary, or "main sequence," star. One of the most important characteristics of this star is that it has a magnetic field. Figure 8.1 shows two views of the Sun made with a ground-based telescope. Figure 8.1(b) shows the Sun as seen in visible light: the disk looks nearly featureless except for a few dark areas, the sunspots. Sunspots are relatively cool areas on the Sun with very strong magnetic fields, fields several thousand times stronger than the Earth's magnetic field. Figure 8.1(a) was made with a special instrument called a *magnetograph*, which produces maps of the solar magnetic field, shown here as black and white areas representing opposite polarities.

The solar magnetic field varies with time in an 11-year cycle. One of the simplest and best indicators of this cycle is the variation in the number of sunspots on the solar surface. Figure 8.2 illustrates the systematic rise and fall in the sunspot number over time from 1610 to 1975. (The most recent sunspot maximum was in 1980.)

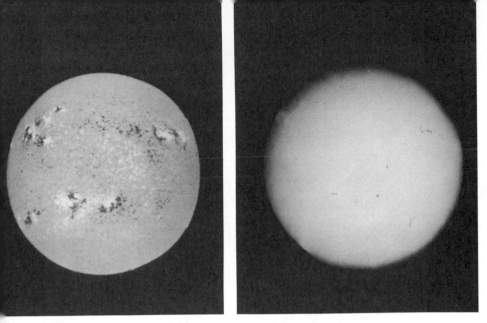

(a) (b)

Figure 8.1: (a) *A picture of the solar magnetic field on the same day; magnetic fields with positive polarity are black, fields with negative polarity are white.* (b) *Photograph of the Sun in visible light; the small dark areas are sunspots.* (Kitt Peak National Observatory photographs)

Note that there was a 60-year period centered around 1670 when there were apparently few sunspots. This is the so-called Maunder Minimum, which happened to coincide with a period of very cold winters here on Earth. (The Pilgrims arrived in New England at the beginning of this period and suffered greatly as a result!) We do not know whether there is a causal connection between the Maunder Minimum (or similar minima that appear to have occurred earlier) and periods of exceptionally cold weather on Earth.

The sunspot number is but one visible result of a solar activity cycle that strongly influences conditions in the Earth's upper atmosphere and ionosphere and the surrounding space. This has practical implications for communications and for the lifetimes and health of orbiting spacecraft. If the activity cycle should prove to have effects on long-term variations in the Earth's lower atmosphere and climate, as suggested by some studies, understanding the sunspot cycle sufficiently to predict future variations would have important economic and political implications.

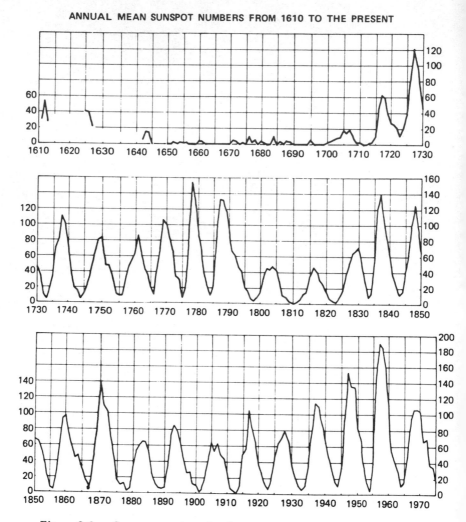

Figure 8.2: *Sunspot numbers for the period 1610 to 1975.* (Courtesy of J. A. Eddy, High Altitude Observatory)

The astronomical significance of the solar activity cycle is less speculative. Many stars apparently have activity cycles similar to the Sun's, cycles that appear to be driven by variations in the stellar magnetic fields. What we do not know is how stars generate magnetic fields. Knowledge about solar and stellar activity cycles may provide important clues to the nature of the dynamos, or "stellar magnetic-field generators," that produce the magnetic fields and cause the activity cycles. The Sun's role is critical because it is the only star whose changing pattern of magnetic fields can be studied across its surface.

Probing the Solar Interior

The surface features of the Sun obviously reflect some processes deep within its interior. Here nuclear fires generate the energy that the Sun pours out into space in the form of radiation, out-flowing gas, and energetic particles. But how can we probe this interior?

One particular type of energetic particle emitted by the Sun, called the *neutrino*, has the unique ability to pass through large amounts of material without being absorbed. Thus, neutrinos produced in the solar core escape from the Sun without being affected by the material they pass through on the way out. By measuring how many neutrinos the Sun emits, one can gain information about the Sun's energy-producing core. Solar neutrinos have been detected by measuring their effects on ordinary cleaning fluid stored in a huge 100,000-gallon vat placed deep in a gold mine shaft to avoid the effects of cosmic rays. Our knowledge about the physics of gases and nuclear-fusion processes allows us to calculate the neutrino flux that might be expected from a star like the Sun. Thus far, we have been getting the wrong answer. The calculated neutrino flux is about three times larger than that measured in the gold mine. This is one of the outstanding problems in solar physics. Has something been left out of the calculations? Or is Mother Nature doing something we have not yet discovered?

To provide possible answers, new methods are been devised to probe the solar interior. These techniques depend upon making use of global vibrations of the Sun. The Sun pulsates, or oscillates, with very small amplitude vibrations that can be detected either as motions on the solar surface or as changes in its brightness. Figure 8.3 shows some of the frequencies with which the Sun oscillates. By studying which frequencies are present and how strong they are, astronomers can probe the solar interior in the same way that geologists probe the Earth's interior using motions caused by

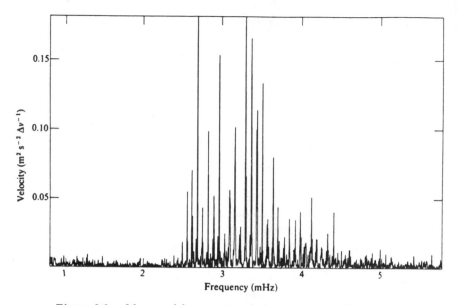

Figure 8.3: *Measured frequencies of global solar oscillations. (A megahertz is 0.001 cycle per second. The mean period of the oscillations is
approximately 5 minutes, which corresponds to a frequency of little more
than 3 megahertz; hence, these oscillations are often called the 5-minute
oscillations.)* (Courtesy of M. A. Pomerantz, Bartol Research Foundation)

earthquakes or explosions. To obtain detailed, precise measurements of global solar oscillations over a wide range of frequencies,
it is necessary to acquire long sequences of observations uninterrupted by clouds, night, or motions of the terrestrial atmosphere.
For example, the measurements seen in Figure 8.3 were obtained
during a five-day period of continuous sunlight during summer at
the South Pole. To get the ultimate observations, appropriate instruments must be placed in space. For example, by the early 1990s,
scientists hope to monitor solar oscillations over a period of months,
even years, using instruments on a European spacecraft positioned
in the L1 libration position between the Earth and Sun. (The L1
point is a position 1.7 million kilometers from Earth, where a
spacecraft remains fixed along the Earth–Sun line as the Earth
revolves around the Sun during the year.) Improved measurements
of solar oscillations will advance our understanding of stellar in-

teriors and perhaps explain why the Sun emits fewer neutrinos than expected.

Probing the Solar Atmosphere

We can also gain information about the Sun by studying its outer envelope, or the solar atmosphere. Figure 8.4 shows the temperature of the solar atmosphere as a function of distance from the surface. At the center of the Sun, where hydrogen is burned to make helium by nuclear fusion, the temperature is very high, 15 million degrees Celsius. The temperature decreases steadily with increasing distance, falling to approximately 4,000 degrees Celsius

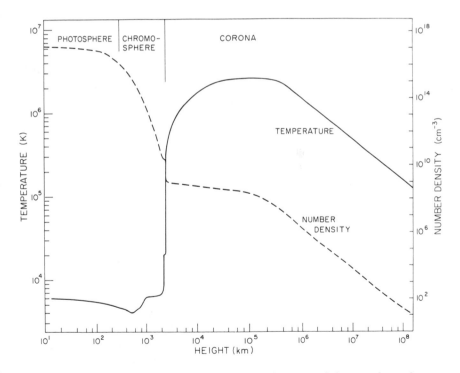

Figure 8.4: *Temperature and density as a function of distance from the solar surface. (Temperatures are given in degrees Kelvin, a scale equal to degrees Celsius plus 273.)*

a few hundred kilometers above the surface. The "surface," known as the *photosphere*, is the layer of the atmosphere we see in visible light with our eyes. Beyond the photosphere are two layers of gas that are transparent to visible radiation, the *chromosphere* and the *corona*. The chromosphere is only a few thousand kilometers thick (thin compared to the Sun's radius of 700,000 kilometers) and has a mean temperature of about 10,000 degrees Celsius. Extending above the chromosphere and far out into space—in fact, out beyond the orbits of the Earth and other planets—is the corona, a tenuous, low-density gas with a mean temperature of about 1.5 million degrees Celsius in the region near the Sun.

One of the outstanding puzzles about the Sun is why the chromosphere and corona are so hot. One would expect the solar atmosphere to become cooler with its increasing distance from the core where nuclear fires are burning, just as the air gets cooler with increasing distance from a hot stove. For many years, scientists thought the unusual heating pattern of the Sun's outer layers was caused by the dissipation of energy carried by sound waves. Just below the solar surface, the gas circulates like boiling water in a kettle, and this convection process produces sound waves that propagate outward. These sound waves dump their energy in the outer atmosphere, and if energetic enough, they could heat up the atmosphere to produce the hot chromosphere and corona. Unfortunately, an experiment on the unmanned Orbiting Solar Observatory (OSO-8) found that although upward-propagating sound waves apparently carry enough energy to heat the lower chromosphere, they do not carry sufficient energy to higher levels to heat the upper chromosphere and corona.

Fortunately, at about the same time, important new measurements of the Sun were acquired by experiments flown on Skylab, the first U.S. Space Station, which was occupied by three crews of astronauts in 1973 and 1974. The Skylab observations provided important clues to the mysterious heating mechanism. Figure 8.5 is an X-ray photograph of the Sun taken by a telescope on Skylab. The brightest areas are regions where substantial coronal heating has resulted in the emission of intense X-ray radiation. These areas are "active regions," with strong magnetic fields and sunspots, and the sites of explosive transient phenomena such as flares. The cooler areas with less X-ray emission are regions with weak magnetic fields.

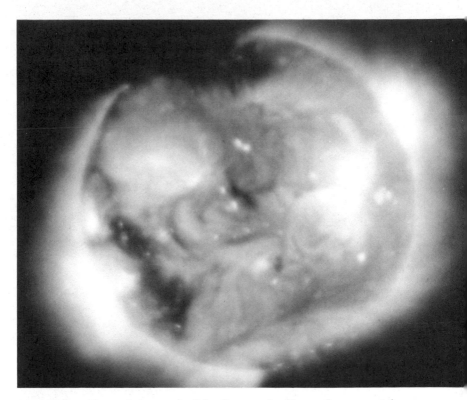

Figure 8.5: *X-ray photograph of the Sun acquired by a telescope on the Skylab satellite. North is at the top, east to the left. The brightest areas are active regions, the dark areas at the north pole and east side are coronal holes.* (Courtesy of *American Science & Engineering* and Harvard College Observatory)

If one places a magnet under a paper covered with iron filings, the iron filings line up to map the magnetic lines of force between the magnet's north and south poles. A hot ionized gas does the same thing. Thus, in the bright regions of the corona, X-ray-emitting gas is confined or "bottled up" by the solar magnetic field and follows the magnetic lines of force. This is illustrated clearly in Figure 8.6, which has several images of an active region photographed in different wavelengths of ultraviolet light. The photograph in the lower right is a picture of the magnetic field similar to Figure 8.1, that is, the white and black areas mark regions of north and south magnetic polarities. The other three photographs

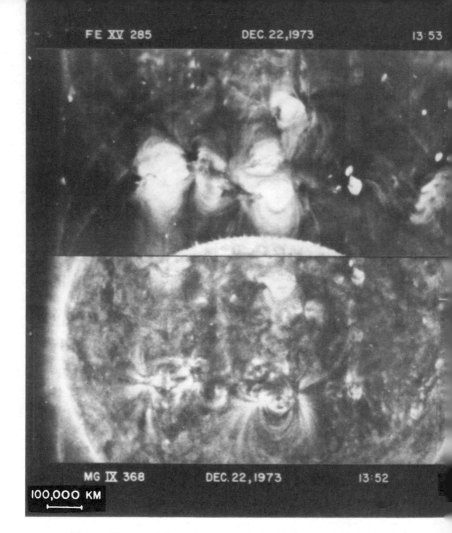

Figure 8.6: Picture of the solar magnetic field (lower right-hand photograph) and photograph of solar active regions in ultraviolet light from gas at different temperatures. The photo labeled Fe XV 285 is from gas at 2.5 million degrees Celsius, Mg IX is from gas at 1.1 million degrees Celsius, and Ne VII is from gas at 500,000 degrees Celsius. These photos show how the high-temperature gas on the Sun delineates the magnetic-field lines between regions of positive (dark) and negative (white) polarity at the solar surface. (Courtesy of the Naval Research Laboratory and Kitt Peak National Observatory)

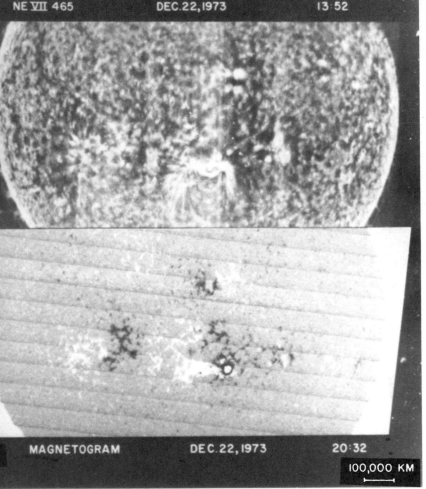

NE VII 465   DEC.22,1973   13:52

MAGNETOGRAM   DEC.22,1973   20:32

100,000 KM

show how the coronal gas lines up along the magnetic-field lines.

The magnetic fields on the Sun vary with time, sometimes rapidly. This variation may produce sudden changes in coronal heating, as illustrated in Figure 8.7. Here a small point of intense X-ray emission (a "coronal bright point") suddenly becomes much brighter and then fades. (Incidentally, this bright point is about the size of the Earth.)

Skylab ultraviolet and X-ray observations such as those shown in Figures 8.5–8.7 suggested that there is an intimate connection between the solar magnetic fields and coronal heating. In fact, we now suspect that most, if not all, of the corona's heating is caused

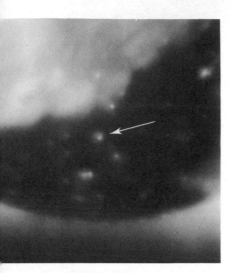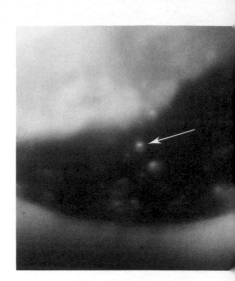

11 JUNE 1973
21:22 UT

12 JUNE 1973
02:01 UT

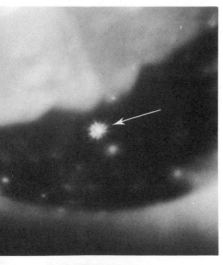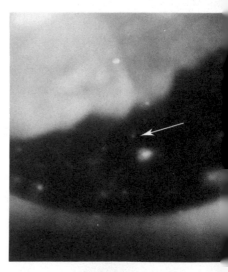

12 JUNE 1973
05:10 UT

12 JUNE 1973
06:46 UT

5 arc min.

by the conversion of magnetic energy into kinetic energy. Continued studies of Skylab X-ray and ultraviolet observations, as well as more recent measurements made with ground-based and rocket instruments, have further supported the hypothesis. However, to understand better the specific heating mechanisms and to establish beyond doubt that the heating is due to dissipation of magnetically stored energy requires much more sensitive instruments.

Another interesting coronal feature prominent in solar X-ray photographs are the large, dark areas known as *coronal holes*. (Examples of coronal holes are visible at the North Pole and left side of the Sun in Figure 8.5.) Coronal holes are regions where the coronal gas is blowing way from the Sun at high speeds (500 to 800 kilometers per second). This gas flows out into interplanetary space to form the "solar wind." Since the Sun rotates once every 28 days, these streams of high-speed solar wind spiral out from the Sun like water streams from a rotating lawn sprinkler. When these streams encounter the Earth's magnetic field, they sometimes jostle it sufficiently to cause energetic particles stored in the Van Allen belt to fall into the upper terrestrial atmosphere near the poles and produce aurorae, the Northern and Southern Lights.

In fact, the Sun controls conditions in the Earth's magnetosphere primarily through the interaction of the solar wind with the Earth's magnetic field. (The magnetosphere is the volume of space dominated by the Earth's magnetism: a dynamic region of interacting low-density gases, magnetic fields, and electric currents within 10 to 20 Earth-radii of the Earth.) The solar wind was first detected by a satellite experiment in 1961. Since then, numerous spacecraft experiments in far-Earth orbit, or in orbits around the Sun, have made in situ measurements of the solar wind at distances as close as 0.3 astronomical unit (AU) from the Sun. (An astronomical unit is the average distance from Earth to Sun, or about 93 million miles.)

Still much remains to be known about the solar wind, because until recently we had no reliable methods for studying the solar

Figure 8.7: *Series of X-ray pictures showing the "flare," or rapid brightening, of an X-ray "bright point." This bright point is about as large as the Earth.* (Courtesy of *American Science & Engineering* and Harvard College Observatory)

wind at its origin, the corona. Except for coronal holes, we have been unable to determine where the solar wind originates on the Sun. Neither do we know what mechanism drives the solar wind away from the Sun. We do know that some of the energy required to accelerate the wind (so that it can escape the Sun's gravitational pull) must be provided by the same unknown mechanism that heats up the coronal gas to its million-degree temperature. In fact, the coronal gas is so hot that its high temperature forces it to flow away from the Sun in all regions where it is not held back or confined by solar magnetic fields. We also suspect that "ripples" in the magnetic-field lines, known as *Alfven waves*, help accelerate or push the solar wind away from the Sun, because the wind flows out at higher velocity than might be expected if it were driven only by the heating mechanism. Alfven waves are motions similar to those produced in a plucked string, that is, waves that "wiggle" the magnetic-field lines.

Measurements of phenomena in the outer solar atmosphere normally can be made only during a solar eclipse, when the Moon covers the bright disk of the Sun so that the much fainter (by a factor of a million) light from the corona can be photographed (Figure 8.8). Natural eclipses occur infrequently, last only a few minutes, and cannot be observed in ultraviolet light (one of the primary emissions from the solar wind) from the ground. However, solar scientists have had considerable success in observing the Sun's outer layers by sending special types of optical instruments known as coronagraphs above the Earth's atmosphere in rockets and satellites. These coronagraphs create artificial eclipses of the Sun in visible and ultraviolet light. Simultaneous observations in visible and ultraviolet light can help provide measurements of the temperature, density, and velocity of the solar wind in various regions of the corona, as well as help determine where and how the solar wind is accelerated.

The study of the solar wind with coronagraphs illustrates the role that different generations of space experiments play in gradually increasing our understanding of a given phenomenon. A 1970 sounding-rocket experiment to photograph the Sun in different wavelengths of ultraviolet light during a solar eclipse first detected ultraviolet emissions from the outer corona, emissions that later could be used to study the flow of the solar wind close to the Sun. This led, in turn, to the development of a new type of coronagraph

Figure 8.8: *Photograph of the solar corona, obtained during an eclipse of the Sun in 1973. The bright regions are called streamers; the dark area over the north (top) pole of the Sun is a large coronal hole.* (High Altitude Observatory photograph)

to study the ultraviolet light emitted by the solar wind. This ultraviolet coronagraph (with a companion white light coronagraph) was flown in 1979, 1980, and 1982 on short-range, suborbital sounding rockets. To gain more observing time than the few minutes afforded by rocket flights, modified versions of the coronagraphs will be flown on several Space Shuttle flights as part of the Spartan

program. The Spartan package, including instrument, power battery, data recording and storage system, and pointing system to
aim the instruments at the Sun (or other astronomical objects), is
carried into orbit by the Shuttle and placed overboard by the
Remote Manipulator System (the Canadian-built "arm"). After a
few days of operation as an independent small satellite in orbit
near the Shuttle, Spartan is picked up and returned to Earth. The
Spartan package provides a relatively low-cost means of testing
experiments in a space environment, while providing much longer
observing times than can be achieved with sounding rockets. The
next stage in coronagraph development will be an instrument flown
on a long-duration unmanned satellite, where several years of operation can be achieved. Ultimately, a large, sophisticated, high-
resolution instrument, the pinhole/occulter facility, will be test-
flown on several Shuttle missions and then mounted on the Space
Station as part of the Advanced Solar Observatory.

Eruptive Prominences and Coronal Mass Ejections

Among the most spectacular forms of solar activity seen by astronomers are the eruptive prominences and coronal mass ejections (Figure 8.9). Prominences are clouds of relatively cool material (about 7,000 degrees Celsius) formed high above the solar
surface and surrounded by a less dense, but much hotter (~1 to
2 million degrees Celsius) coronal gas. The prominence material
is suspended in the corona by magnetic-field lines; but, occasionally, these magnetic forces become unbalanced and they drive the
prominence material outward away from the Sun at velocities of
up to hundreds of kilometers per second. The photograph in Figure
8.9 of a large eruptive prominence was acquired 19 December
1973, by the Skylab astronauts, shortly after it began accelerating
outward. (I remember vividly this particular event because I was
working with the Skylab Mission Control team in Houston and
was asked by the Flight Director to give the press conference for
the day. The primary item of scientific interest for that day's conference was, of course, this extraordinary prominence that had
erupted at a perfect time for the astronauts to make a variety of
measurements and photographs. Indeed, this photograph became
one of the most widely distributed pictures made during the Skylab
mission.)   .

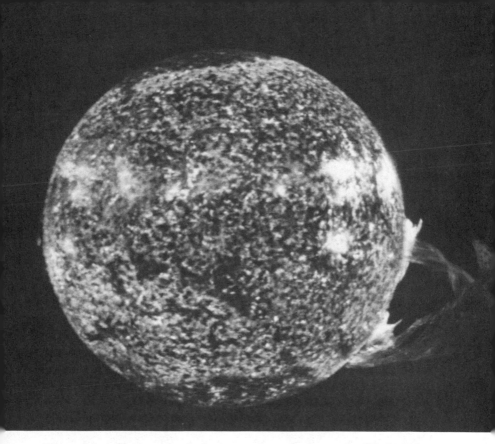

Figure 8.9: *Photograph of a large eruptive prominence that lifted off the Sun on 19 December 1973. At the time of the photograph, the prominence extended over 250,000 miles above the solar surface. The mass of this prominence was about 4 billion tons.* (Naval Research Laboratory photograph)

While that photograph shows only the cooler material being blown off the Sun by magnetic forces, Figure 8.10 illustrates what happens to the hot coronal gas in front of such an eruption. The coronal material piles up, often forming a shock wave, much like the bow wave in front of a boat, and moves out into interplanetary space. If this material happens to be ejected in the direction of the Earth, it, too, perturbs the Earth's magnetic field and, like the solar wind, may cause energetic particles from the Van Allen belt to flow down the Earth's magnetic-field lines at the poles to create aurorae.

Experiments on satellites flown in the past decade have shown

233:13:33

14:41

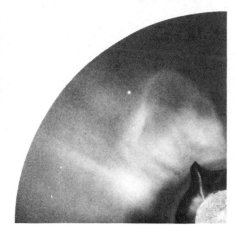

15:11

16:15

Figure 8.10: *Composite pictures of an eruptive prominence and coronal transient, photographed by Skylab instruments on 21 August 1973. The solar disk and prominence were photographed in ultraviolet light from helium at chromospheric temperatures. The surrounding coronal gas was photographed using a coronagraph that made an artificial eclipse so that faint white-light emission from the corona could be seen. As the prominence erupted from the solar surface at 14:41 (2:41 P.M.), a cloud of coronal gas was blown away from the Sun ahead of the prominence, as can be seen in the photographs from 14:11 to 16:53 (2:11 to 4:53 P.M.).* (Courtesy of the High Altitude Observatory and Naval Research Laboratory)

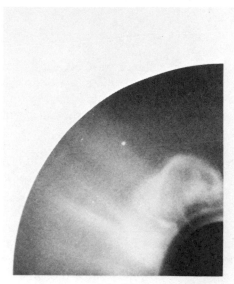

14:50

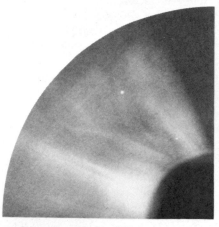

16:53

that coronal mass ejections occur approximately once or twice per day. In fact, this was one of the unexpected discoveries made by Skylab's solar experiments. Although we have learned a great deal about these mass ejections, including their frequency, their association with eruptive prominences and solar flares, and the amount of mass ejected, many questions remain unanswered. Why do mass ejections occur? What causes the sudden unbalancing of magnetic forces to drive them? Is the ejected material heated by the shock wave? How—and where—does the shock wave accelerate atomic particles to high energy? The answers to these questions will require new observations, particularly observations made with high spatial resolution in the coronal layer close to the Sun's surface where the mass ejections originate.

Flares

Solar flares are small areas that get very hot (10 to 20 million degrees Celsius) in a few minutes. Flares appear to be caused by the sudden release of large amounts of energy stored in the magnetic fields of an active region. As this energy is released, it is converted into high-energy particles (electrons and protons) that collide with the gas in the magnetic loops, or arches, of the active region, heating it to high temperatures and producing intense ultraviolet, X-ray, and gamma-ray radiation. Frequently, prominence and coronal material near the site of a flare will be violently ejected out from the Sun deep into interplanetary space. Flares come in a wide range of sizes and energies: some produce large fluxes of energetic particles, others very small fluxes.

The higher the temperature of a gas, the more electrons are stripped from the atoms by collisions with other atoms and electrons in the gas. Atoms with electrons removed are called *ions*, and each atom and ion has unique wavelengths of radiation. By taking pictures at the appropriate wavelength, then, one can photograph gas at a particular temperature. Actually, most of the radiation produced by the ions of high-temperature gases are found in the far ultraviolet and X-ray parts of the spectrum; hence, our interest in acquisition of observations from these regions.

Again, the Skylab mission proved extremely valuable because of its ability to observe in different wavelengths in the far ultraviolet. The top two photographs in Figure 8.11 show flare gas at temperatures of about 2 million degrees Celsius, with the obser-

Figure 8.11: *Extreme ultraviolet pictures of a large solar flare showing its appearance in gas at temperatures of about 2 million degrees Celsius (Mg × λ625), 300,000 degrees Celsius (O VI λ1032), and 20,000 degrees Celsius (C II λ1335). (Harvard College Observatory photograph)*

7 September 1973 FLARE

1411 UT   1545 UT

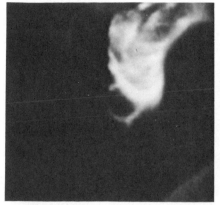 

Mg x λ 625   Mg x λ 625

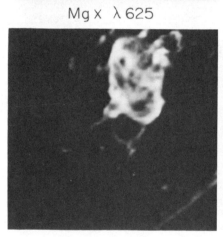 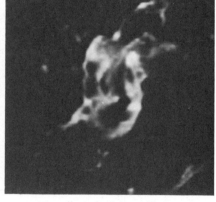

O vi λ 1032   O vi λ 1032

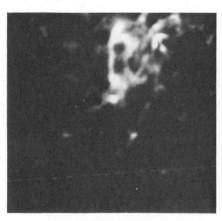 

C ii λ 1335   C ii λ 1335

vations separated by about 1.5 hours. This particular flare was large enough to last several hours

Another important result of space observations is the finding that the intense X-ray and ultraviolet radiations from flares originate in "magnetic arches," or great hollow tubes filled with high-temperature gas. One can see several of these arches, looking much like the famous McDonald's golden arches, in the Mg X images of Figure 8.11. The "footpoints" of these arches are rooted in surface regions of strong photospheric magnetic fields; for example, in the C II images on the bottom row, the left-hand set of footpoints are regions of bright chromospheric emission. (C II is the designation of a carbon atom from which one electron has been stripped; C I is neutral carbon. C II radiation originates in the chromosphere. Similarly, Mg X is a magnesium atom from which 9 electrons have been stripped.)

Much of our current knowledge about the physics of flares has been deduced from observations from spacecraft such as Skylab, as well as from the SMM and Japanese *Hinotori* satellites. The SMM, or Solar Maximum Mission satellite, which was successfully repaired by Shuttle astronauts in April 1984, has been particularly effective in probing flares with a battery of instruments specifically designed for the coordinated study of these phenomena. To make further progress, however, requires new instruments that can probe the fine structure of flares and the active regions from which they originate. While all the instruments flown thus far have provided a wealth of information about these interesting phenomena, they have had insufficient resolving power to "see" how the magnetic arches in active regions evolve as they store up energy. Most likely, the energy is stored in twists of the magnetic-field lines, much like energy is stored by the twists in a rubber band. But questions remain: are the twists already in the field lines as they emerge from below the solar surface; or are the twists introduced by movements of the magnetic-arch footpoints? By studying movements of the footpoints at photospheric levels, and the resulting shapes of the loops, it may be possible to understand what conditions cause the magnetic arches to become unstable and suddenly "unwind," thus releasing stored energy as a flare.

Because of "shimmering" in the apparent image of the solar disk produced by motions in the Earth's atmosphere, it is not possible to observe the magnetic fields with sufficient detail using ground-based telescopes. It is necessary to use a large telescope in orbit, and the proposed Solar Optical Telescope (SOT), which operates in the visible and near ultraviolet region of the spectrum, along with other high-resolution instruments operating in the far ultraviolet and X-ray regimes, should provide the data critical for understanding the physics of the Sun—and how flares work.

The Next Generation of Solar Telescopes in Space

One advantage of making solar observations from space is illustrated in Figure 8.12, which shows several images of a region near the edge of the Sun in different ultraviolet wavelengths corresponding to different temperatures. The top row of pictures shows, from left to right, images of gas at temperatures of 20,000 (Lyα), 100,000 (C III), and 300,000 degrees Celsius (O VI); while the bottom row, from left, shows gases at 500,000 (Ne VII), 1.5 million (Mg X), and 2.5 million degrees Celsius (Fe XV). Note that as the temperature increases, the pictures are less "sharp" or show less detail. This is because structures in the solar atmosphere have more fine detail and smaller dimensions in the lower, cooler levels than in the corona, where the temperature exceeds 1 million degrees Celsius. From the ground, we can only obtain photographs of the chromosphere that appear similar to the ultraviolet image in the upper left (although with better spatial resolution). To see the real three-dimensional character of the solar atmosphere, we need to observe in colors of light that cannot get through the Earth's atmosphere. In addition, as mentioned earlier, we need to see the fine details of the solar structure, particularly the subtle interactions between magnetic fields and gas. This requires not only telescopes that must be in space, but that must be large enough to resolve the fine details of the solar atmosphere.

An example of the type of fine structure that cannot be resolved because of atmospheric "seeing" is shown in Figure 8.13. This is a "quiet" region of the solar disk, that is, one without active regions or sunspots. Made when the Earth's atmosphere was very steady and thus did not shimmer very much, this image records the narrow band of red light emitted by hydrogen gas in the solar chromosphere, that thin layer of 10,000-degree-Celsius gas just above

ACTIVE REGION ON EAST LIMB

30 JUNE, 1973

Ly α   C III   O VI

Ne VII   Mg X   Fe XV

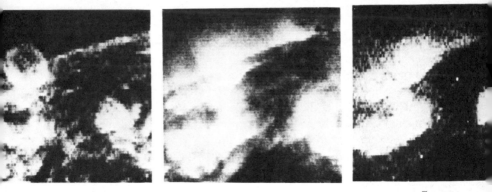

Figure 8.12: *Pictures of the solar limb, showing several active regions as seen in ultraviolet light from gas at different temperatures. From left to right in the top row, 20,000 degrees Celsius (Lyman alpha), 100,000 degrees Celsius (C III), and 300,000 degrees Celsius (O VI) and, in the bottom row, 500,000 degrees Celsius (Ne VII), 1.5 million degrees Celsius (Mg X), and 2.5 million degrees Celsius (Fe XV).* (Harvard College Observatory photographs)

the solar surface. The dark rows are jets of material called *spicules*, which form the fundamental fine structure of the chromosphere. To resolve these features, we need observations with spatial resolution 10 times better than now is normally possible from telescopes on the ground. And until we can do this, we are in the position of someone trying to figure out what a forest is without being able to get a close-up of a tree.

The Solar Optical Telescope will be able to resolve the fine structure of the Sun's surface (photosphere) and its lower atmosphere (chromosphere). This large telescope (approximately 7 meters long with a mirror 1.25 meters in diameter) is designed to be operated from the Space Shuttle or, eventually, from a more permanent platform, such as the Space Station. Figure 8.14 is an artist's conception of SOT operating in orbit from the Shuttle, with SOT the large, barrel-like structure resembling an old-fashioned siege mortar in the aft end of the Shuttle bay. The large-diameter hole in the forward end of the barrel is the entrance aperture of the 1.25-meter telescope; the three smaller holes are the entrance apertures of other co-observing instruments used to make measurements in far-ultraviolet and X-ray light. (Instruments operating in the far-ultraviolet and X-ray bands of the electromagnetic spectrum need different types of telescopic mirrors.) We expect these telescopes to be flown on several Shuttle missions in the early 1990s.

The structure with the long boom mounted in the forward end of the Shuttle bay in Figure 8.14 is a second type of solar facility: the pinhole/occulter facility. The boom is used to position a flat disk, or plate, at a large distance (50 meters) from several types of detectors located in the Shuttle bay (on the mounting structure just behind the Shuttle cabin in this Figure). The plate on the end of the boom has two functions. It casts a large shadow, thus creating an artificial solar eclipse so the optical instruments operating in the ultraviolet and visible parts of the spectrum may make high-resolution observations of coronal radiation extending from the solar surface out to large distances. The plate is also perforated with "pinholes" that let through hard X-ray radiation. In fact, combined with detectors in the Shuttle bay, the facility acts as a giant pinhole camera to make high-resolution pictures of hard X-ray emission from the Sun. (It is not possible to observe hard X

rays with conventional reflecting telescopes since the rays pass through, or are absorbed by, the mirrors.)

High-resolution hard X-ray images of the Sun are needed for studying the acceleration of high-energy particles in the solar atmosphere, particularly in flares. Energetic high-speed electrons produce hard X-ray radiation when they pass through the solar atmosphere. Hence, by taking pictures of the Sun in rapid time sequences, we can trace the paths of high-energy electrons, measure their acceleration, and determine how and where they give up their energy as they collide with the chromospheric and coronal gases. Understanding the physics of particle acceleration in the solar atmosphere is one of the primary objectives of solar astronomers, and observations with the pinhole/occulter facility should provide critical new observations on this fundamental problem.

The Solar Optical Telescope, the pinhole/occulter facility, and the co-observing instruments mounted in the SOT canister form the basic components of the Advanced Solar Observatory, a battery of instruments designed to probe the Sun from a long-duration space platform. Initially these instruments will be flown on Shuttle sortie missions, both to check out the instruments and to obtain selected observations of various solar features and phenomena. No doubt, these early observations will lead to discoveries that will stimulate the development of even more improved detection systems. Eventually, however, scientists hope to operate the telescopes over long periods, ranging from months to years, so they can observe the evolution of long-lived phenomena, such as active regions and coronal holes, with typical lifetimes of weeks to months. We want to learn, for example, why active regions are more prolific in the production of flares when they are young than when they are old. We also want to know what determines the energy output of a flare and how this energy is divided between radiation, en-

Figure 8.13: *Photograph of the chromosphere near the Sun's edge, obtained in red hydrogen light using the tower telescope at Sacramento Peak Observatory and taken at a time when the Earth's atmosphere was very steady, thus providing excellent "seeing." The dark rows are jets of material called* spicules, *which make up the chromospheric network. These jets are about 10,000 kilometers long and 1,000 kilometers wide; they shoot up with velocities of about 25 kilometers per second and remain visible for 5 to 10 minutes.* (Sacramento Peak Observatory photograph)

Figure 8.14: *An artist's conception of a Shuttle-borne Solar Optical Telescope (the large, barrel-shaped instrument mounted in the aft end of the Shuttle bay) and pinhole/occulter facility (the group of instruments in the forward end of the Shuttle bay behind the 50-meter-long boom).* (NASA/Marshall Space Flight Center illustration)

ergetic particles (cosmic rays), and coronal mass ejections. In fact, the Sun is one of the few places in the universe where we can watch the rapid conversion of magnetic energy into a wide variety of other forms. To understand fully this energy transmutation, it is necessary to "catch" a number of flares and observe them from beginning to end. Moreover, we must monitor in detail their birthplaces, particularly before the flares occur, so we can determine where and how the flare energy is stored and what triggers its explosive release. This type of study obviously requires long pe-

Figure 8.15: An artist's conception of the Advanced Solar Observatory, a battery of solar instruments mounted on a long-duration platform such as the Space Station. Along with the Solar Optical Telescope and pinhole/occulter facility are other instruments, including high-energy gamma-ray detectors and several solar radio telescopes whose antennas are shown mounted at various positions on the platform. (NASA/Marshall Space Flight Center illustration)

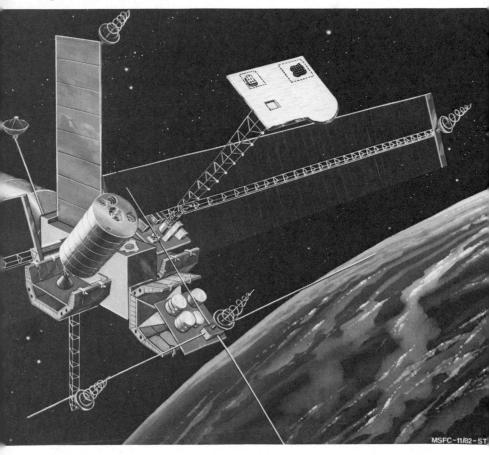

riods of observations possible only from a long-duration platform such as the Space Station.

An artist's concept of one possible configuration of future solar instruments arrayed on a space platform is shown in Figure 8.15. This complex of instruments, the Advanced Solar Observatory, will have unprecedented capabilities for probing the Sun's secrets and, one hopes, for providing the data necessary for understanding the physics of this very special star—our Sun.

AXAF

A Permanent X-Ray Observatory in Space

MARTIN V. ZOMBECK

We have always gazed at the stars. Ancient astronomer/astrologers watched the heavens, counting and cataloging the stars, noting the seasonal changes in their positions, navigating by them, and linking the lives of men with their motions. In the modern era, scientists have become more analytical perhaps, seeking to define the physical nature of the stars rather than their effects on the fortunes of men. Yet, on the larger scale, the attempts to measure the size of the universe and to understand the processes and mechanisms that underlie its nature ultimately lead to questions about the origin and final fate of the universe—and life. Understandably, then, not only scientists but theologians, philosophers, and poets have concerned themselves with cosmological questions.

Of course, while poets and philosophers may ponder the mysteries of the universe, only astronomers directly take its measure. They do this by observing the direction, quantity, and quality of the radiation reaching their eyes or their instruments from celestial objects. Amazingly, until just after World War II, or little more than 40 years ago, all that was known about the universe was the result of observations in a very narrow region of the spectrum: the band visible to the human eye and a small portion of radiation to either side in the near infrared and near ultraviolet. With this very restricted viewpoint, the image of the cosmos was one of serenity, of stability, and, except for an occasional exploding star or supernova, of modest changes taking place over incredibly long time intervals.

This picture of a "quiet universe" began to change dramatically after 1945. Radio astronomy, born out of wartime radar and communications instrumentation, opened a new window to reveal the universe as the seat of explosive events leading to extremely high-temperature gases, of particles accelerated to very high energies, and of extreme conditions of density and pressure. Active galaxies were found emitting electromagnetic radiation at rates a million times greater than normal galaxies like our own. Strange objects known as quasars, apparently not much larger than our Solar System, were found to radiate more energy than several hundred galaxies combined. Supernova remnants, the gaseous debris of cataclysmic stellar explosions, were found to shine as the result of gases being raised to million-degree temperatures or from the emissions of high-energy electrons spiraling in magnetic fields. The relic radiation of the Big Bang was found, thus supporting the theory that the known universe was created from a tremendous explosion some 10 to 20 billion years ago. Thus, throughout the universe nature has conspired to produce conditions ideal for the production of high-energy radiation, especially X rays, which are found in copious amounts whenever gases are heated to millions of degrees or when violent events accelerate matter to high velocities.

Not surprisingly, then, the field of X-ray astronomy, which began just a little over 20 years ago, has grown into a mature discipline, ranking with radio and optical astronomy as essential for the study of the universe. This chapter offers a brief history of this field, discusses some of its remarkable discoveries, and describes a future tool for X-ray astronomy, the Advanced X-ray Astrophysics Facility, or AXAF. AXAF will be a semipermanent X-ray observatory in space and, together with the Hubble Space Telescope in the optical region and the ground-based Very Long Baseline Array (VLBA) in the radio region, will form a triad of instruments for fundamental studies in astronomy and astrophysics into the twenty-first century.

Why, we might ask, is the field of X-ray astronomy so young if the universe seems to reveal a major aspect of its nature through the emission of X rays? The obvious answer is that we cannot "see" X rays. Our basic "optical instrument," the human eye, is sensitive only to a very narrow band of the electromagnetic spectrum and has no response to X rays or, for that matter, ultraviolet, infrared, gamma-ray, and radio radiation.

Technically, X rays are a form of "light," but like the other forms of electromagnetic radiation outside the visible band, X rays are described by their wavelength or by the energy of their associated light particle or photon rather than by a subjective attribute such as "color." The unit of wavelength is the angstrom (Å), which has a length of one-hundred-millionth of a centimeter; the unit of photon energy is the electronvolt (eV). The shorter the wavelength of the radiation, the more energetic the photon. For comparison, green light has an average wavelength of about 5,000 angstroms, and its photon has an energy of about 2.5 electronvolts; whereas X rays of wavelength 1 angstrom have photon energies of 12,400 electronvolts. Obviously, X rays are thousands of times more energetic than visible light.

Ever since the 1895 discovery of X rays in the laboratory by Wilhelm Roentgen, we have been developing instrumentation to convert X rays into visual images or into electronic pulses that can be counted. Fluorescent screens, special photographic films, and Geiger counters are all examples of devices for recording X rays from terrestrial sources. Recording celestial sources, alas, is more complicated. The Earth's atmosphere prevents us from seeing X-ray radiation in the heavens, even while it simultaneously protects us from these and other cosmic rays. (The chart in the introduction to this book illustrates the attenuation of celestial radiation due to the ocean of air above us.) For example, in the X-ray band between 100 angstroms and 0.1 angstrom (124 electronvolts, and 124,000 electronvolts), only 1 percent of the radiation striking the top of the Earth's atmosphere penetrates to altitudes between 20 and 100 kilometers. At an X-ray wavelength of 1 angstrom (12,400 electronvolts), the Earth's atmosphere is equivalent to a shield of lead approximately 1½ inches thick. In short, X-ray emissions from celestial objects are completely inaccessible to instruments on the Earth's surface. Hence, X-ray astronomy could not exist until it

was possible to place instruments at altitudes high above the atmosphere; vehicles for accomplishing this became available
only after the dawn of the Space Age.

Indeed, X-ray astronomy traces its roots to the postwar rocket
experiments of the late 1940s. Captured German V-2 rockets were
used to loft film and high-energy radiation counters above the
Earth's atmosphere and thereby established the existence of X-
ray emission from the Sun's corona. The observation of the Sun
in X rays continued with more and more sophisticated instruments
until 1973, when the manned Skylab satellite carried X-ray telescopes capable of imaging the hot solar corona with a clarity almost
equal to that of optical telescopes (see the Withbroe chapter in
this book). The NASA program of solar X-ray astronomy spurred
development of many instrumental techniques later used to open
an X-ray window on the universe.

The first detection of a cosmic (i.e., nonsolar) X-ray source, and
thus the true beginning of X-ray astronomy, was accomplished by
a rocket flight in 1962. The primary goal of this experiment had
been to search for X rays produced by the bombardment of the
Moon's surface by energetic atomic particles from the Sun. Lunar
X rays were not detected. However, when the rocket's Geiger
counters scanned in the direction of the constellation Scorpius, an
intense source of X-ray emission was discovered. This source, named
Sco X-1, was later found to be associated with a faint star and to
be emitting 1,000 times more energy in X rays than at visible
wavelengths. Thus, Sco X-1 was an entirely new class of object.
This same rocket flight, which lasted only five minutes, also detected the presence of a diffuse background of X rays across the
entire sky. Again, later observations showed this background to
be remarkably uniform, but its origin is still not understood. Indeed, the whole celestial sphere seems to be aglow in X rays.

For eight years after the discovery of Sco X-1, observations were
carried out with larger and more sophisticated detectors, but still
on rocket flights lasting no more than a few minutes each. About
three dozen X-ray sources were discovered, most apparently within
our galaxy, but a few of extragalactic origin. Because most of these
sources were located to a precision of no more than one-half degree, only a handful of the sources could be identified with known
optical objects.

In December 1970, the first satellite devoted to X-ray astronomy was launched from the Kenya coast in East Africa. This satellite had the official name of Small Astronomy Satellite 1, or SAS-1, but the scientific team responsible for the observatory renamed it *Uhuru* ("freedom" in Swahili) since the satellite was launched on Kenya's Independence Day. During its two-year life span, *Uhuru* performed an all-sky survey in X rays and detected over 300 sources of X-ray emission.

Figure 9.1 shows a map of the sources detected by *Uhuru*, with the apparent X-ray brightness of each source indicated by the size of the dot. The coordinate system is the standard galactic system used by astronomers. The horizontal line about which most of the sources are clustered is the galactic equator, or Milky Way. Some familiar—and not so familiar—objects are here: Sco X-1, the first—and still the brightest—X-ray source to be discovered; the Crab nebula, the remnant of a supernova observed by the Chinese in the year 1054; M31, or Andromeda, our twin galaxy; the large and small Magellanic Clouds, neighboring galaxies in the southern sky; 3C273, a quasar; and Cygnus X-1, the best candidate for a black hole. (Only relative brightnesses are shown on this map; thus Andromeda, which is actually about 200 times more luminous in X rays than Sco X-1, appears relatively faint because it is 1,000 times farther away.)

One major scientific result of *Uhuru* was the discovery that a large fraction of the galactic X-ray sources detected are binary systems in which matter from a normal star is falling onto an extremely dense object, such as a white dwarf, a neutron star, or, as we believe in at least one case, a black hole. These binary systems emit 1,000 times more energy in X rays alone than the Sun radiates in all wavelengths together.

A second major result was the discovery of X-ray emission from high-temperature gas in clusters of galaxies, the largest aggregates of matter in the universe, in the region between the galaxies. This 100-million-degree gas, although invisible to optical telescopes, appears to have as much mass as the galaxies embedded within it.

A progression of U.S. and European satellites followed *Uhuru*, culminating with the launch in November 1978 of the High Energy Astronomical Observatory 2 or, as it was better known, the *Einstein Observatory*. Einstein carried the first telescope capable of detecting and imaging X rays from celestial objects other than the Sun. Besides providing fine details of many objects with an angular

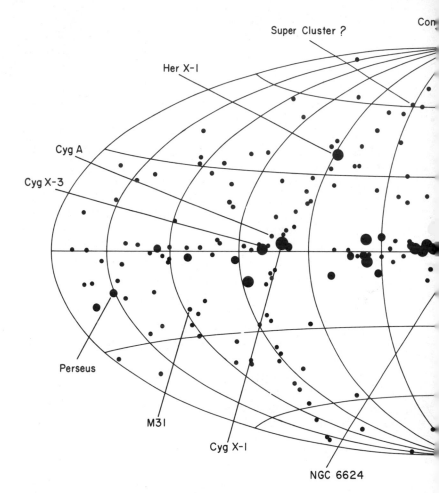

Con

Super Cluster ?

Her X-1

Cyg A

Cyg X-3

Perseus

M31

Cyg X-1

NGC 6624

Figure 9.1: *A map of the sky using galactic coordinates, showing the locations of over 300 sources detected by the* Uhuru *X-ray satellite. The Milky Way runs from left to right about the horizontal line, and the center of our galaxy appears in the middle of the diagram.* (Smithsonian Astrophysical Observatory illustration)

resolution of a few arcseconds, its greatly increased sensitivity extended X-ray observations to almost every type of astronomical object in the universe. In fact, almost overnight the imaging ca-

RU CATALOG

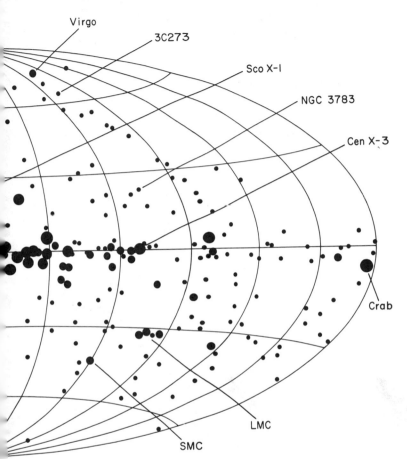

pabilities of Einstein made X-ray astronomy a major subdiscipline of modern astronomy.

The imaging telescope of the Einstein Observatory is rather unconventional when compared with traditional optical instruments. As might be expected, X rays will simply penetrate the surface of a normal mirror without reflection. However, X rays can be reflected if they strike surfaces at large angles of incidence; in other words, if they can be made to bounce like a stone skipped over a pond. Thus, X-ray mirrors are not flat, dishlike plates, but rather cylindrical glass tubes whose inner surfaces have been shaped and polished. This type of collector is called a *grazing incidence telescope*. Incoming X rays simply graze the curved mirror surfaces and are brought to a focus some distance away. The focused X rays are then converted into electronic pulses, which can be transmitted by radio to the ground, where they are processed and reconstructed into television images. To increase the collecting area of such a telescope, several concentric cylinders are nested, one inside the other. Figure 9.2 illustrates the concept of the grazing

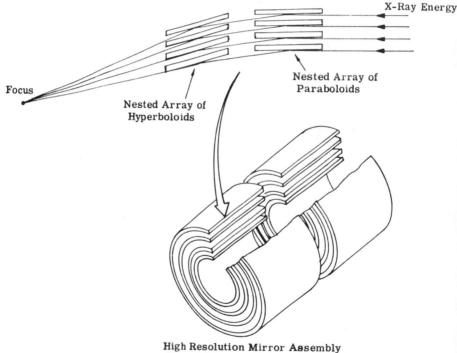

Figure 9.2: *A schematic drawing of the Einstein Observatory grazing incidence mirror assembly. X rays from a distant object graze the nearly cylindrical surfaces and are brought to a focus, where they are detected by a variety of imaging and spectroscopic instruments.* (Smithsonian Astrophysical Observatory illustration)

incidence telescope. The Einstein Observatory's assembly of mirrors had a 0.6-meter-diameter aperture and a 3.4-meter focal length. It was about 1,000 times more sensitive than the *Uhuru* X-ray satellite.

Figures 9.3 and 9.4 illustrate some of the wide variety of astrophysical phenomena observed by Einstein, representing X-ray emission from all types of known astronomical objects. For example, all classes of stars—from dwarfs to supergiants—were found to emit X rays. Additional X-ray pulsars have been discovered embedded in supernova remnants. And an X-ray jet up to 300 light-years in length has been seen emanating from the central object in a strange system in our galaxy known as SS433.

At much larger distances beyond the Milky Way, more than 100 individual X-ray sources have been observed in our twin galaxy, Andromeda. X-ray jets have been detected in several active galaxies, including the radio galaxy Centaurus A, shown in Figure 9.4. X-ray structures more than one million light-years in diameter have been mapped from the hot gas in distant clusters of galaxies. X-ray emission has been detected from faint quasars at the very edge of the visible universe. And closer to home, X rays have been observed from the planet Jupiter (Figure 9.5).

Many supernova remnants were imaged in X rays by the Einstein Observatory. Figure 9.6 shows how the supernova remnant Cassiopeia A (Cas A) appears in X rays. As the result of a stellar explosion, a blast wave carrying the outer envelope of the destroyed star expands into the interstellar medium, thereby heating the gas to about 10 million degrees so that it glows brightly in X rays. From the X-ray map of Cas A, we can estimate the mass of the original star; it is about 15 times the mass of our own star, the Sun.

Stunning as the Einstein observations were, they amount to no more than a mere glimpse of the X-ray universe. We have barely tapped the vast resources of X-ray astronomy for investigating the nature of the cosmos. We are now ready for the next step: from pioneering observations of relatively short duration to a long-term, highly sensitive, X-ray observatory equipped with the most advanced instruments available. The Advanced X-ray Astrophysics Facility (AXAF) is the natural extension of the past two decades of research.

The AXAF will be a semipermanent observatory placed into

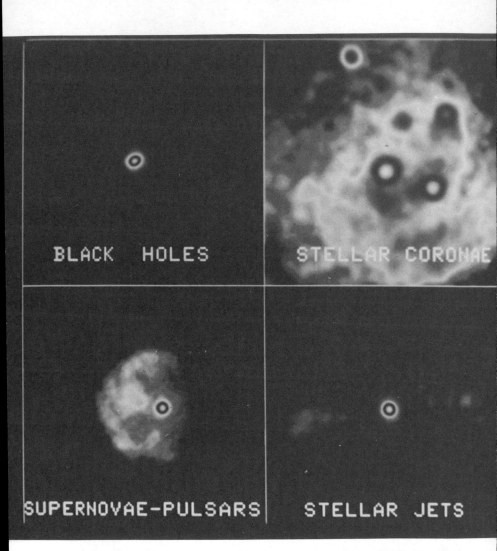

Figure 9.3: *X-ray images of objects in our galaxy obtained by the Einstein Observatory. Clockwise from top left: Cygnus X-1, Eta Carina nebula, CTB 109, and SS433.* (Smithsonian Astrophysical Observatory photograph)

Earth orbit by the Space Shuttle. The eye of the AXAF is a large-area, high-resolution, grazing incidence mirror assembly similar to that of Einstein but comprised of six nested pairs of mirrors instead

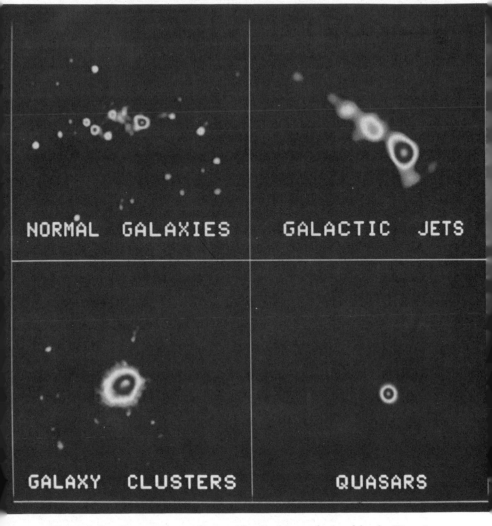

Figure 9.4: *X-ray images of extragalactic objects obtained by the Einstein Observatory. Clockwise from top left: Andromeda galaxy, Centaurus A, 3C273, and Abell 2256.* (Smithsonian Astrophysical Observatory photograph)

of four (Figure 9.7). The diameter of the aperture is 1.2 meters and the focal length is 10 meters. AXAF is designed to operate for 10 to 15 years, and during its lifetime, it will be considered a

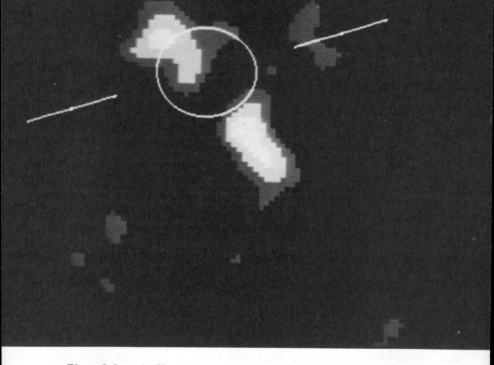

Figure 9.5: *An X-ray image of the planet Jupiter obtained by the Einstein Observatory. An outline of the planet's disk and the location of its equator are shown. The X-ray emission occurs high in Jupiter's atmosphere.* (Smithsonian Astrophysical Observatory photograph)

major national observatory in space, to be used by the entire international astronomical community. The detectors at its focal plane will provide for imaging, spectroscopy, and polarization studies with a precision never before achieved in X-ray astronomy.

Figure 9.6: *An X-ray image of the supernova remnant Cassiopeia A (Cas A) obtained by the Einstein Observatory. The shell of 10-million-degree gas and the stellar debris is still expanding 300 years after the initial stellar explosion.* (Smithsonian Astrophysical Observatory photograph)

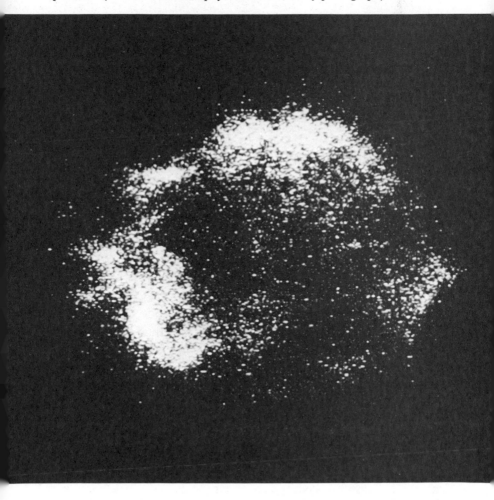

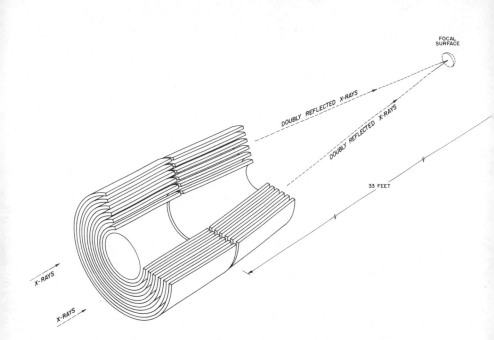

Figure 9.7: *A schematic drawing showing the six nested pairs of grazing incidence mirrors of the AXAF telescope. The focal length of the assembly is 10 meters (33 feet), and the outer diameter is 1.2 meters (4 feet). The mirrors can focus X rays with energies of up to 10,000 electronvolts. An X-ray-emitting object with an angular diameter equivalent to that of the Moon (one-half degree) will produce an image with a diameter of 90 millimeters (3.5 inches). Normal galaxies similar to our galaxy will be detected at distances up to 1 billion light-years with this X-ray "eye."* (Smithsonian Astrophysical Observatory photograph)

The scientific data will be transmitted to the ground for processing and analysis, using the tracking and data-relay satellite system (TDRSS). During its life, the AXAF will be visited by the Shuttle astronauts for routine maintenance, repair, and the replacement of instruments. The latter capability is especially important, for it will allow the AXAF to keep pace with advances in detector technology.

The AXAF will have a sensitivity some 100 times that of the Einstein Observatory. As a result, it should be able to detect the coronal X-ray emissions of some 100,000 stars in our galaxy, supernova remnants in neighboring galaxies, and quasars more remote than the most distant now observed in any wavelength, and to map with precision the 100-million-degree gas in clusters of galaxies.

In addition to increased sensitivity, AXAF's mirror assembly will have 10 times better angular resolution than does the Einstein Observatory. Not only will we see more detail, but new and unknown sources previously lost in diffuse haze will begin to appear. For example, the Einstein Observatory was able to detect some 100 individual sources in the nearby galaxy Andromeda; AXAF should be able to extend this detection of X-ray sources in other galaxies to all 2,500 members of the Virgo cluster of galaxies. (The Virgo cluster is 60 million light-years away, whereas the Andromeda galaxy is only 2 million light-years distant.)

Figure 9.8 is an artist's concept of the AXAF, which will be about 50 feet long and weigh about 10 tons. Figure 9.9 shows a schematic diagram of the AXAF and identifies its major components. At least a half-dozen individual focal-plane detectors will be alternately placed into the focus of the telescope on command. They include X-ray cameras to provide fine details of the shape and structure of X-ray sources, spectrometers to provide information on the chemical composition and temperature of these objects, and polarimeters to provide details of the mechanisms for generating X rays. Every three or four years, the Space Shuttle will rendezvous with the AXAF for routine maintenance and supply and to install new and improved focal-plane instrumentation.

Although the grazing incidence mirrors bring X rays to the focus of the telescope, rather novel instruments are necessary to detect the image. One of these is an X-ray image intensifier based upon a microchannel plate (MCP). This device is capable of detecting a single X-ray photon and providing an electronic signal that yields an accurate position of the photon within an image. A microchannel is a small-diameter (ranging from 0.0005 to 0.001 inch) glass tube, across which an electrical potential of about 1,000 volts is applied. When an X ray strikes the wall of the tube, it knocks out an electron. This electron is accelerated by the electrical field within the tube, whereupon it strikes the wall again, producing two or three more electrons. These electrons are accelerated in turn, thereby producing more electrons. The process repeats and repeats again until the original electron is multiplied some 1,000

Figure 9.8: *An artist's concept of the Advanced X-ray Astrophysics Facility (AXAF). The AXAF will be approximately 50 feet long and weigh about 10 tons. Launched by the Space Shuttle and serviced and supplied while in orbit, AXAF is expected to operate for 10 to 15 years, thereby making it the first "permanent" X-ray observatory.* (Courtesy of Richard Redden, Lockheed Missile and Space Company)

to 10,000 times. Thus a veritable cloud of electrons with sufficient magnitude to produce a measurable electrical signal emerges from the rear of the microchannel tube (Figure 9.10). Typically, 10 million of these tubes can be fused together to form an MCP

(Figure 9.11). This entire array operates as an image intensifier, converting the X-ray image into an amplified electronic image that can then be displayed on a television screen.

One X-ray-imaging instrument under development is based upon the solid-state electronic detector, called a CCD (charged-coupled device), now widely used as television-camera sensors. A product of the microelectronic revolution, the typical CCD is a 1-inch-square silicon chip containing about 1 million accurately positioned photosensors. An X-ray photon incident upon one of the photosensor cells releases electrons within the cell, with the number of

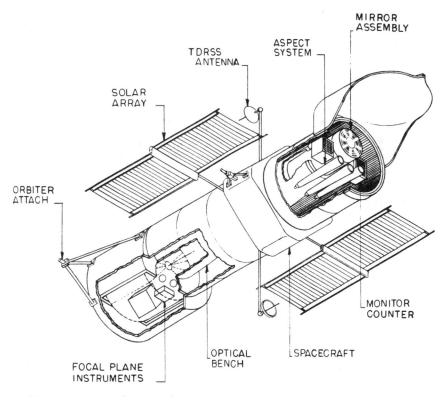

Figure 9.9: *A schematic drawing of the AXAF showing the major subsystems, including the mirror assembly, array of focal-plane detectors, and the solar-power array.* (Smithsonian Astrophysical Observatory illustration)

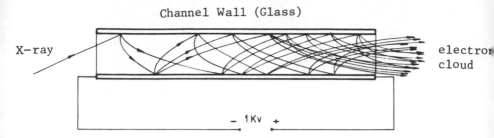

Figure 9.10: *X-ray conversion and electron multiplication in a micro-channel.* (Hammamatsu Corporation illustration)

released electrons proportional to the energy of the arriving X ray. By electronic means, it is possible to measure both the magnitude of the released electronic charge and the position within the array of the particular photosensor that detected the X ray. Thus, both the energy and position of an individual X-ray photon can be recorded and transmitted to a video screen for immediate display or to a computer for storage and later analysis. This means that "multicolor" images can be made of X-ray sources and, after analysis, temperature, gas pressure, chemical composition, and density maps can be derived from the same data. (As David Latham has noted [page 27], CCDs are now also being used extensively in visible and infrared astronomy.)

The Einstein Observatory stopped operating in the spring of 1981, when the on-board gas used to position the satellite was exhausted; it burned up upon reentry into the Earth's atmosphere a year later. Two X-ray satellites are now operating, the European Space Agency's EXOSAT and Japan's TENMA, both launched in 1983. EXOSAT has two small imaging X-ray telescopes that respond to "soft" X rays (energies less than 2,000 electron volts), but both have poorer spatial resolution than the Einstein telescope and a combined collecting area only about one-quarter that of Einstein. TENMA has no X-ray-imaging capability. Both satellites have limited lifetimes of two to three years.

The next X-ray telescope to be launched will be the West German ROSAT (Roentgen satellite). This observatory, with a spatial resolution comparable to Einstein's, is scheduled for launch in late

1987. ROSAT will respond only to soft X rays, has no capability for spectroscopy or polarimetry, and has a lifetime of only two years. NASA is also planning to carry a special X-ray instrument on a series of relatively brief Space Shuttle flights in the late 1980s. This device, known as LAMAR (large-area modular array of reflectors), has moderate spatial resolution (0.5 to 1 arcminute), a large collecting area, and sensitivity to X rays with energies up to 10,000 electronvolts. Still, none of these observatories, either singly or combined, can offer the long-duration capability, high spatial resolution, broad X-ray energy range, high sensitivity, and spectroscopic and polarimetric capabilities of AXAF.

Indeed, the opportunities offered by AXAF for long-term detailed and systematic studies of high-energy processes in the universe are unique. Using the AXAF, we expect findings related to stellar and cosmic evolution, the structure and dynamics of celestial objects, and perhaps clues to the ultimate fate of the universe. Indeed, many of the questions about the "hidden matter" in the

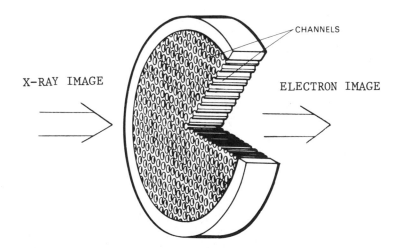

Figure 9.11: *Schematic of a microchannel plate (MCP), consisting of an array of millions of small-diameter glass tubes (microchannels). The diameter of the plate ranges from 25 to 120 millimeters, and the thickness is about 1 millimeter. The MCP converts an X-ray image into an amplified electronic image.* (Hammamatsu Corporation illustration)

universe seem linked to high-energy phenomena visible only in X rays. This ambitious NASA program is already under way: major design studies have started and (at least at the time of this writing!) launch is planned for 1992. This launch date represents a 10-year pause in the United States' efforts to advance the field of X-ray astronomy. The launch of the AXAF is not assured, however. Before actual construction can begin, the program must be approved by the U.S. Congress. Scientists are optimistic that this will happen and that one of the most exciting windows on the universe will be opened wide again.

The Gamma-Ray Universe

Taking a Long, Hard Look

TREVOR C. WEEKES ·

The universe, when glimpsed through different bands of the electromagnetic spectrum, presents many contrasting views. And the gamma-ray astronomer's view may be the most exciting of all.

The optical astronomer sees a sky essentially populated by stable objects. Stars and galaxies, many of which have exhibited no change over millions of years, are the major components. The Sun is a typical inhabitant of this reassuring, if somewhat dull, universe.

At shorter wavelengths, X-ray and ultraviolet astronomers begin to see a much more dynamic and unstable universe, in which change is the rule rather than the exception. But the gamma-ray astronomer sees a universe that is violent, chaotic, and explosive. It is a unique view of the hottest, most energetic regions of the cosmos. This view is also unique in that gamma-ray astronomy is an infant science and therefore the picture is only dimly seen.

Gamma ray is the generic term applied to the rather extensive band of photons seen at the shortest wavelengths (and thus the highest energies) of the electromagnetic spectrum. This term was first applied to the penetrating and lethal radiation from radioactive atoms. This radiation has energies in the range of 1 to 5 mega-electron volts.* Although the electromagnetic spectrum extends far beyond this, the same term *gamma ray* is used to describe even the photons a million million times more energetic. As might be expected over such a wide range of energies, a wide variety of detectors, or gamma-ray "eyes," must be used to study quite different aspects of each energy band (Figure 10.1).

Gamma-ray photons are almost always secondary to the very-high-energy particles (electrons, protons, and heavier nuclei) that constitute the cosmic radiation. The existence of this radiation has been known since 1912; it constantly bombards the Earth's upper atmosphere (and its secondary products reach sea level), but little is known about where it originates or how such very great energies can be achieved in space.

In its most basic interpretation, then, gamma-ray astronomy can be seen as the search for an understanding of the cosmic radiation—its point of origin, its extent, its relation to other cosmic phenomena, its mechanism of acceleration, and its propagation through the interstellar medium.

Because the cosmic particles, being charged, lose their original

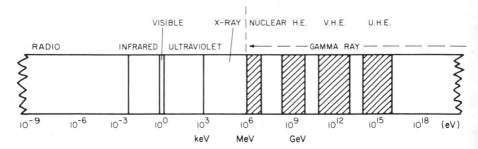

Figure 10.1: *The electromagnetic spectrum from the perspective of the gamma-ray astronomer.* (Smithsonian Astrophysical Observatory illustration by John Hamwey)

* The convenient unit of energy in high-energy astrophysics is the electronvolt (eV). For reference, the energy of a blue photon of light is 3 electronvolts. The megaelectronvolt (MeV) is one million electronvolts. The gigaelectronvolt (GeV) is one billion electronvolts. The rest mass of the proton is equivalent to 1 gigaelectronvolts.

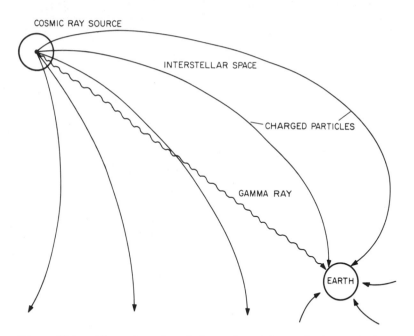

COSMIC RAY SOURCE

INTERSTELLAR SPACE

CHARGED PARTICLES

GAMMA RAY

EARTH

Figure 10.2: *The trajectories of charged particles and photons in inter-stellar space. Photons travel in straight lines; charged particles are bent by magnetic fields and appear to arrive at the Earth in equal quantities from all directions.* (Smithsonian Astrophysical Observatory illustration by John Hamwey)

directionality, gamma rays are the most direct tracers of their activity (Figure 10.2). The presence of high-energy particles can be inferred in some celestial objects by their radio, optical, and X-ray emission, but only gamma rays unambiguously point to the presence of these particles. Moreover, while many high-energy processes are too energetic to be duplicated in a man-made laboratory, the study of gamma rays permits a direct investigation of the energy processes at the heart of some of the universe's most unusual high-energy sources.

A quick look around the heavens with gamma-ray eyes soon reveals a view that is quite different from that seen by the optical astronomer. The Solar System is barely apparent; the planets are invisible, and the Sun is detectable only during violent solar flares. The gamma-ray sky is dominated by the Milky Way; not, however, as a vast conglomeration of stars, but by *the space between the stars*—interstellar space. The cosmic rays that fill the galaxy move freely here, but they occasionally collide with the weak concentration of hydrogen that is distributed throughout interstellar space. These interactions produce penetrating gamma rays that trace out the gas, as well as the cosmic-ray concentrations, and give us a new probe of galactic structure.

Scattered along the galactic plane are discrete (individually distinct) sources of gamma rays, the most prominent of which are the Crab and Vela pulsars, the remnants of supernova explosions that took place hundreds of years ago. Since pulsars (rapidly rotating neutron stars) are widely believed to be very efficient particle accelerators, it is no surprise that they should also be gamma-ray sources. In fact, the detection of gamma rays provides direct evidence of high-energy particle activity. Other discrete gamma-ray sources are associated with molecular clouds, but the majority have not yet been identified.

Perhaps the most interesting gamma-ray source is the mysterious object known as Cygnus X-3. First discovered by X-ray astronomers in 1967, Cygnus X-3 came to the attention of the whole astronomical community in 1972 when it suddenly flared to become the brightest radio source in the sky. Subsequently, it was shown to be a gamma-ray source detectable over a vast range of gamma-ray energies.

More surprising, the high-energy emission from Cygnus X-3 is not constant; rather, it shows regular periodic variations, such as those seen in many binary X-ray sources. The repetition rate of these variations is 4.8 hours, but the pattern of the variations changes with energy (Figure 10.3). The energy source in Cygnus X-3 is not known; it may involve a pulsar or a black hole, but this is still only speculation. Most certainly it is a source of very high-energy particles. But Cygnus X-3 may also be the most powerful cosmic-ray source in the galaxy.

Away from the plane of the galaxy, the gamma-ray sky is relatively uniform; this broad, diffuse, and relatively feeble flux probably originates far outside the galaxy. The monotony of this region of the sky is broken by a few sources beyond the galaxy. The quasar 3C273 and the radio galaxy Centaurus A are dimly seen, showing that gamma-ray emission comes from beyond the galaxy as well. It is still an open question as to whether or not the bulk

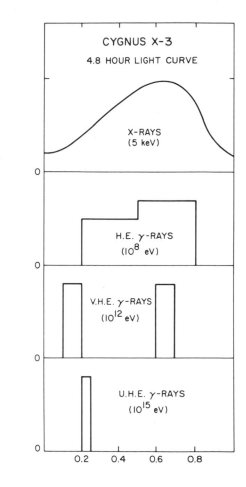

Figure 10.3: *The variation of Cygnus X-3 with time; the output at various gamma-ray energies is folded at the 4.8-hour period of the binary source. The character of the light curve changes dramatically as we go to higher energies.* (Smithsonian Astrophysical Observatory illustration by John Hamwey)

of the cosmic radiation in our vicinity originates in the galaxy or beyond it.

Despite the unique insight that gamma-ray studies can provide, the gamma-ray sky is still virtually unexplored. The reason is quite simple: gamma-ray astronomy is *hard*—not only in the sense that it involves the study of hard, penetrating photons, but also hard in the sense that it is difficult. Gamma-ray photons at all energies are relatively sparse. More important, perhaps, while these photons can cross millions of miles of intergalactic and interstellar space, they cannot penetrate the Earth's atmosphere. In fact, the atmosphere has the equivalent absorbing power of a lead wall 1 meter thick. This is reassuring in terms of its shielding power for human beings, but very frustrating for the Earth-bound gamma-ray astronomer.

Like X rays, gamma rays can be detected directly only from above the Earth's atmosphere. Unlike X rays, however, gamma rays can be detected only with difficulty. At very high energies, cosmic-ray particles and gamma-ray photons behave similarly. Unfortunately, because the charged cosmic rays are 100 to 1,000 times more numerous, they complicate the gamma-ray observations. There is a certain irony here: the origin of the cosmic radiation is the major motivation to do gamma-ray astronomy, but this same cosmic radiation makes the study very difficult!

Gamma-ray detectors must be very cleverly designed to capture gamma rays while at the same time rejecting the much more numerous electrons, protons, and heavier nuclei of the cosmic radiation. Luckily, nature has been kind enough to provide a gamma-ray interaction (gamma-ray photon + matter) that has a very characteristic signature at energies above 10 megaelectronvolts. A gamma-ray photon, being uncharged, will pass through the outer walls of a typical detector without leaving an ionizing trail. However, if the gamma-ray photon passes close to the nucleus of an atom in the thick target, the photon disappears and a pair of electrons takes its place. Moreover, the electrons share the energy of the photon and retain much of its original trajectory. And the electrons, being charged, leave trails of ions in the detector. The gamma-ray signature is thus that of a nonionizing particle producing two ionizing particles (Figure 10.4).

The practical gamma-ray detector, then, needs some method of making those ionizing tracks visible. The electron pair appearing, as it were, out of nowhere can then be identified and the trajectory of the original gamma-ray traced back into space. In the 100-megaelectronvolt region, the best detector of choice is the spark chamber (Figure 10.5), in which the ionizing paths of the electrons are made visible by a trail of sparks. (The chamber is surrounded by a scintillator shield that responds to the charged particles but not to gamma rays, and thus acts as a veto.) The trail of sparks can be recorded visually or electronically, and the resultant pictures scanned by a trained observer, or by computer, to identify and measure gamma-ray events. Typically, such a device can determine the arrival directions of gamma rays from space to within 2 degrees.

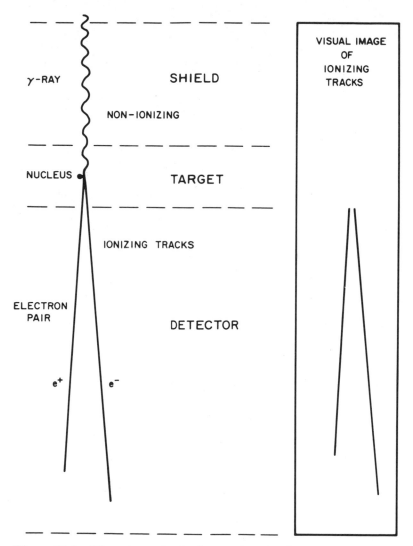

Figure 10.4: *The pair-production interaction: the photon produces a pair of electrons (positive and negative) in the field of a nucleus; the interaction can be seen in a detector that makes the ionizing trails visible.* (Smithsonian Astrophysical Observatory illustration by John Hamwey)

In fact, most of our knowledge of the gamma-ray sky has come from two remarkably successful spark-chamber satellite experiments, the United States' SAS-2 mission launched in 1972 and the European COS-B mission launched in 1975. Although the effective collection area of each of these spark chambers on these instruments was only 60 square centimeters, they provided the first maps of gamma-ray intensity across the sky and established the broad features of the gamma-ray sky: the diffuse background, the galactic plane emission, and the existence of some 30 discrete sources. The position of these objects was known to an accuracy of 1 to 2 degrees, too large to permit identification with known optical objects based on position alone. Only when a source exhibited time variations at a characteristic period (e.g., the Crab and Vela pulsars), or where there was a prominent object in an otherwise "empty"

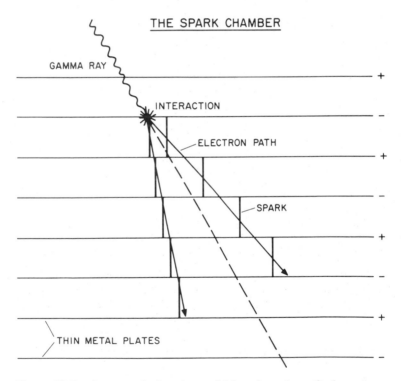

Figure 10.5: In a spark chamber, a high voltage is applied to alternate plates separated by an inert gas. The passage of a charged particle through the plates and gas leaves a trail of ions that causes a spark to jump between the plates. The sparks can easily be recorded with an optical recorder. (Smithsonian Astrophysical Observatory illustration by John Hamwey)

field (e.g., 3C273 or Rho Ophiuchus), could identifications be made with certainty.

The nature of the other sources remains a mystery. Indeed, perhaps the most significant aspect of these sources is that they have *not* been identified. Most of the brightest gamma-ray sources in the sky are not objects whose behavior at lower energies makes them particularly noteworthy. Thus, there is the exciting prospect that a whole new class of objects may be generating high-energy particles by hitherto unknown processes.

New instruments with better angular resolution and increased sensitivity may reveal many other weaker sources. But detection of new sources, although significant, really tells us little about the physical nature of the objects. Their positions must be well determined so that they can be optically identified. A detailed study of their energy spectrum and time variability must be made to provide vital clues to the gamma-ray emission mechanism.

The next gamma-ray satellite experiment is a joint Soviet–French venture known as GAMMA-1. It is similar in concept to SAS-2 but has 2.5 times the collection area and somewhat better angular and energy resolution. GAMMA-1 will undoubtedly confirm and extend the SAS-2 and COS-B pictures of the gamma-ray sky; however, the next major advance must await the launch of the Gamma Ray Observatory (GRO) late in this decade. This joint European–United States mission will actually involve four completely separate, but complementary, experiments. One will be an all-sky, low-energy monitor, designed to detect and locate "gamma-ray bursters," low-energy gamma-ray sources that flare brightly for perhaps a minute and then disappear. These gamma-ray bursts are currently detected by an array of small satellites at the rate of 10 per year; with GRO, the rate should increase to hundreds per year.

The second GRO experiment is a spectrometer designed to detect gamma-ray spectral lines that result from radioactive decays on the Sun and other stellar sources. (Optical spectral lines correspond to changes in the atom; gamma-ray spectral lines come from the nucleus.) Although GRO itself will move slowly on a predetermined observing program to survey the entire sky in one year, the spectrometer can be offset to study transient sources such as solar flares or supernova explosions.

The third instrument covers the very difficult energy region around

30 megaelectronvolts. At this energy level, the production of a pair of electrons is not the dominant interaction of gamma rays in the detector; instead, the gamma-ray photon collides with an electron and transfers most of its energy to it. Thus it is not so easy to positively identify the gamma rays. Although the sensitive area of the GRO spectrometer will be less than 50 square centimeters, it will be a factor of 10 more sensitive than any previous experiment.

Perhaps the most interesting experiment on GRO is the energetic gamma-ray experiment telescope (EGRET), the direct successor of the SAS-2 and COS-B spark-chamber experiments. A joint venture of the Goddard Space Flight Center, Stanford University, the Max Planck Institut in Germany, and the Grumman Aerospace Corporation, this detector consists of spark chambers surrounded by an anticoincidence detector and a large sodium iodide crystal to detect the emerging electrons (Figure 10.6). Its important features will be its greater collection area (2,000 square centimeters), better energy resolution, and improved angular resolution (one-third of a degree). It will also cover a very wide energy range (20 megaelectronvolts to 30 gigaelectronvolts).

GRO will be launched from the Space Shuttle. It will then be moved to a higher orbit to avoid excessive aerodynamic drag that would pull it back to Earth. However, this orbit will not be high enough to enter the South Atlantic Anomaly, a region with a large concentration of trapped charged particles that would adversely affect the observatory's performance. The initial mission is scheduled to last two years, but the observatory could be refurbished by Shuttle astronauts and its life extended.

The history of astronomy has shown that dramatic new discoveries result whenever new instruments increase sensitivity by a factor of 10. Since GRO will achieve a sensitivity improvement of 10 to 20 over a wide range of energies, knowledge of the high-energy universe should advance significantly within the coming decade.

The upper energy limit of gamma rays observable with EGRET is 30 gigaelectronvolts, partly because of the limited instrument size and partly because of the sparsity of gamma rays at this very high energy. Even if the detector had a collection area as large as one square meter, it would see only one gamma ray per week above 10 gigaelectronvolts from the strongest known source. Of

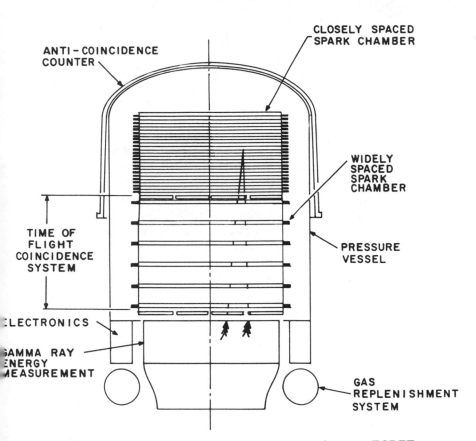

ANTI-COINCIDENCE
COUNTER

CLOSELY SPACED
SPARK CHAMBER

WIDELY
SPACED
SPARK
CHAMBER

TIME OF
FLIGHT
COINCIDENCE
SYSTEM

PRESSURE
VESSEL

ELECTRONICS

GAMMA RAY
ENERGY
MEASUREMENT

GAS
REPLENISHMENT
SYSTEM

Figure 10.6: *The high-energy spark-chamber experiment known as EGRET,*
to be flown on the Gamma-Ray Observatory (GRO) in 1988. (NASA
illustration)

course, the electromagnetic spectrum extends to energies far above
10 gigaelectronvolts. In many ways, this end of the spectrum is the
most interesting since these gamma rays involve physical processes
of such incredible energy that they cannot be duplicated in any
man-made accelerators.

Imagine how limited our knowledge of the visible universe would
be if we did not have telescopes and we had to rely on detector
size (the human eye) to define our collection area for light rays.
This is the situation in gamma-ray astronomy. Because gamma
rays cannot be reflected or refracted like optical photons, there is
no easy way to increase the collection area of a space-borne gamma-
ray detector. Eventually, when Space Station experiments become

a reality, it may be feasible to assemble a massive array of gamma-ray detectors with collection areas of, say, 100 square meters. But such arrays are not likely in this century, so we must turn to other, rather indirect, methods of detecting very-high-energy gamma rays.

These methods predate the Space Age, but they have come into prominence only with the increased interest in gamma-ray astronomy as a result of the lower energy satellite experiments. As shown in Figure 10.4, the interaction of a very-high-energy gamma ray with matter produces an electron pair. If the energy is sufficient, the resulting electron pair does not quickly lose its energy by ionizing interactions. Instead, an electron in the field of a nucleus can produce a secondary gamma ray. This process is called *bremsstrahlung*, from the German word for "braking radiation." In brief, the electron brakes suddenly in the vicinity of the nucleus, slows down, and emits a gamma ray. This gamma ray can then produce another electron pair that can undergo further bremsstrahlung interactions. The result is an electromagnetic cascade (photons, electrons, and positrons), which continues traveling in the original direction of the primary gamma ray and sharing its total energy.

Because the Earth's atmosphere is equivalent in its absorbing power to a shield of lead 1 meter thick, it can serve as an effective medium for the development of an electromagnetic cascade. Typically, the gamma-ray photon undergoes its first interaction at an altitude of 20 kilometers. The atmosphere is thin at this altitude, so the particles in the resulting cascade have time to spread out before undergoing further interactions, thus creating an "air shower" that effectively *magnifies* the gamma ray by a factor of 1,000 to 10,000.

An observer at sea level (Figure 10.7) sees the air shower as a pancake of particles (mostly electrons and positrons) with a diameter of 100 meters and a thickness of 1 meter. These charged particles can easily be detected with Geiger counters or scintillators. Nor is it necessary to detect *all* the particles. A few detectors spread out on the ground and tied together to respond in coincidence can sample parts of the pancake; the arrival direction can be determined from the difference in arrival times at the separate detectors. Similarly, the energy of the primary gamma ray can be estimated from the particle density measured at each detector. The only complication in this scheme is that the numerous charged cosmic rays also produce cascades of particles. In this case, the

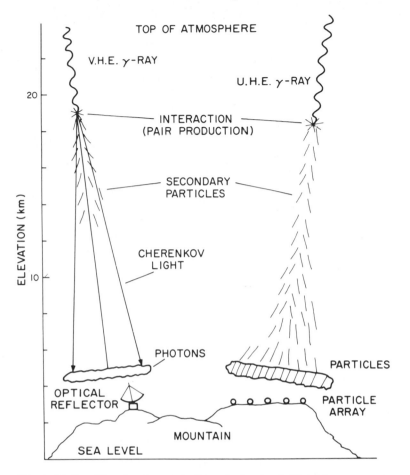

Figure 10.7: *The development of an air shower initiated by gamma rays in the atmosphere. In the case of very-high-energy gamma rays, only optical photons survive to detector level; at ultrahigh energies, the particles survive.* (Smithsonian Astrophysical Observatory illustration by John Hamwey)

particle pancake that reaches ground level contains penetrating particles (muons and nuclei) in addition to the electrons and positrons of the electromagnetic cascade. It is possible to reduce this cosmic-ray background with fairly sophisticated detector arrays, even if it cannot be eliminated entirely.

Many such air-shower arrays have been built, usually to study the very-high-energy interactions in the atmospheric air shower, and it has been possible to detect at least one ultrahigh-energy (1,000,000 gigaelectronvolt) gamma-ray source: Cygnus X-3. The first gamma rays associated with this object were seen by experimenters at the University of Kiel in 1983; the detection was confirmed shortly afterward by a group at the University of Leeds. Since then, researchers at the University of Adelaide have reported detecting Vela X-1 (another binary X-ray source). The magnitude of these achievements can be appreciated when one realizes that the observed signal corresponds to one gamma-ray photon per square meter per 300 years! In practice, of course, the arrays had effective collection areas of more than 1,000 square meters, but it still took three to four years of continuous operation to detect a significant signal.

The energy threshold of these arrays (which are all close to sea level) is 1,000,000 gigaelectronvolts. By going to mountain altitudes, the threshold can be reduced by a factor of 10. However, any air showers produced by photons of less than 100,000 gigaelectronvolts just do not have enough surviving particles to be detected. Fortunately, there are other techniques for detecting the lower-energy photons. In 1934, the Russian physicist Pavel Cherenkov discovered that a relativistic particle passing through a medium in which the velocity of light is less than that of the particle will disturb the medium in such a way that it will emit a feeble blue light: this is called *Cherenkov radiation*. In 1953, Bill Galbraith and John Jelley at Harwell in the United Kingdom demonstrated that air showers cause the emission of optical Cherenkov radiation in the atmosphere. This proved an elegant way to detect gamma rays in the intermediate energy region between those detectable only from satellites and those of ultrahigh energies. In the Earth's atmosphere, the secondary particles cause this Cherenkov radiation, and it arrives at the Earth's surface as a pancake of mostly blue photons with a radius of 100 meters and a thickness of 1 meter. Unlike the secondary particles, however, this light penetrates to ground level, even from a shower whose particles have been absorbed high in the atmosphere. At a dark site, under a cloudless, moonless sky, a simple system of light detectors can easily detect the pulse of light. The angular spread of the light is so small (about 1 degree) that the arrival direction of the gamma ray can be determined with fairly high accuracy.

The great virtue of the atmospheric Cherenkov technique is its simplicity. Optical detectors are widely available and relatively inexpensive. Moreover, the optics can be of fairly poor quality. Indeed, surplus army searchlight mirrors often have been used because they have just the right optical quality. The conversion of the brief optical light pulse (less than 10 billionths of a second) into an electrical signal is accomplished with phototubes. Fortunately, the electronics used in nuclear physics experiments are ideal for this purpose. And, as in the air-shower array detectors, the collection area can be much larger than the detector size.

Once again the major limitation of the technique is the background of charged cosmic rays. There is no easy way to distinguish the light pulse from a gamma-ray air shower from that associated with a proton shower. The arrival directions of the charged particles are thoroughly mixed by the magnetic fields of interstellar space; therefore, the background cosmic rays are isotropic. The gamma-ray source will be apparent only as an excess of showers from the direction of the source (Figure 10.2). In a typical detector, the gamma-ray signal may be only 1 percent of the background. Many hours of observations are therefore required to establish the existence of a source; since the observations can be made only under clear, stable skies, it can take some months to detect a source.

Although optical astronomers and very-high-energy gamma-ray astronomers are interested in cosmic photons whose energies differ by a million million times, the observing requirements of both are similar. For this reason, very-high-energy gamma-ray experiments are generally located at optical observatories.

In the past decade, a small but significant number of very-high-energy gamma-ray sources have been detected by observers in Europe, Asia, North America, and Australia. Because the atmospheric Cherenkov technique depends on the comparison of a candidate source with a background region, none of the detected sources is "new." In other words, there had to be some reason to suspect an object might be a gamma-ray source for it to be studied in the first place. Thus, the sources chosen for investigation usually are those that are very luminous at other wavelengths or that otherwise show evidence of high-energy particle activity.

In our galaxy, the best candidates for high-energy particle activity are the pulsars. Not surprisingly, then, the first very-high-energy gamma-ray source detected was the pulsar in the Crab nebula. Similarly, among the first extragalactic objects to be examined were the bright radio galaxies. None of those studied in the Northern Hemisphere yielded a signal, but Centaurus A in the Southern Hemisphere proved to be a strong source. Gamma rays of 1,000 gigaelectronvolts from Cygnus X-3 were detected by the group at the Crimean Astrophysical Observatory in 1972 and have since been seen at all gamma-ray energies, ranging from 100 megaelectronvolts to 10,000,000 gigaelectronvolts. At all energies, the emission from Cyngus X-3 varies with a period of 4.8 hours, but the form of the variation is different at each energy (Figure 10.3).

Because each of the objects listed above is the brightest member of a distinct class of celestial objects, there must be a rich harvest to be reaped in this energy range. A small improvement in sensitivity should dramatically increase the number of detected sources.

Since 1968, the atmospheric Cherenkov instrument with the lowest effective energy threshold (about 100 gigaelectronvolts) has been the Smithsonian Astrophysical Observatory's 10-meter optical reflector at the Fred Lawrence Whipple Observatory in southern Arizona. From a distance, this instrument looks very much like a radar antenna, but close up (Figure 10.8), it can be seen to consist of 248 small, spherical optical mirrors, each with a focal length of 7.3 meters.

The quality of these mirrors is crude by optical astronomy standards; however, the large collection area (four times that of the 5-meter Mount Palomar telescope!) more than compensates for the angular resolution (about one-third of a degree). Since the Cherenkov light images are typically 1 to 2 degrees in diameter, this poor resolution is not a real limitation. For more than a decade, this instrument, using a single-channel detector with a beamwidth of one degree, was the most sensitive detector in the 100- to 1,000-gigaelectronvolt energy region.

Although this conventional version of the atmospheric Cherenkov detector (a single photometer at the focus of a mirror) has the virtue of simplicity, it does not exploit all the information available in the Cherenkov light image. This kind of detector simply registers an event whenever the light intensity in the phototube exceeds a preset threshold in a time interval of 10 nanoseconds (1 nanosecond = 1 billionth of a second). As early as 1960, it was realized that if a "snapshot" is taken of each Cherenkov light image, it is possible to more accurately determine the arrival di-

Figure 10.8: *The 10-meter optical reflector at the Whipple Observatory in southern Arizona.* (Smithsonian Astrophysical Observatory photograph)

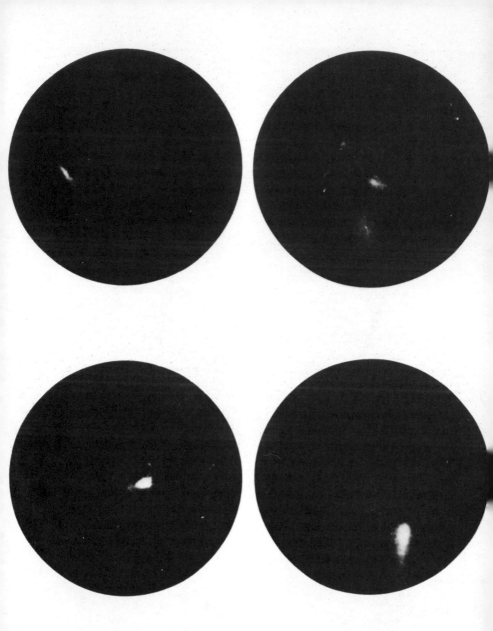

Figure 10.9: *Cherenkov light images from cosmic-ray air showers, as recorded by an image-intensifier camera developed by M.I.T. and University College, Dublin, in 1960.* (Pictures courtesy of J. White)

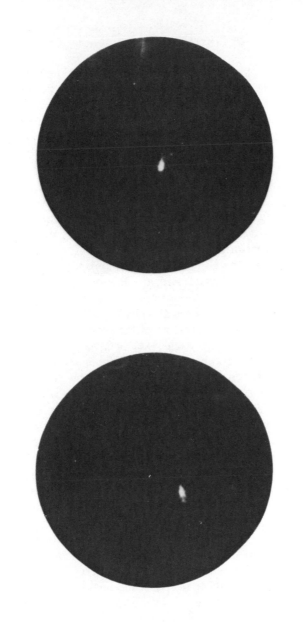

rection of the electromagnetic cascade—and hence the originating gamma ray! Also, by improving the accuracy of arrival-direction determination, the background of charged particles can be reduced and the detection of weaker sources of gamma rays made possible.

Taking a snapshot of a phenomenon one million times faster than the fastest shutter speed on a good 35-millimeter camera is just as difficult as it sounds! The air showers arrive at random times and from all directions. The dark night sky is not really all that dark when you try to detect a weak and diffuse light source against it. Many fast-imaging techniques are available (image intensifiers, video cameras, electronic shutters), but they are small and they cannot be coupled with the poor optics of typical atmospheric Cherenkov systems. The net effect of these limitations is that although Cherenkov light images were successfully recorded more than two decades ago (Figure 10.9), the technique has not been used in gamma-ray astronomy.

In 1977, Ted Turver of the University of Durham and I suggested coupling this "imaging" approach with the low-energy threshold of the 10-meter optical reflector. This could be accomplished by using an array of small phototubes as the pixels (picture elements) of an imaging system at the focal plane of the reflector. One way to explain its operation is that each pixel behaves like a light-sensitive grain on a photographic plate. Phototubes serve well as pixels: they are fast and sensitive, as well as inexpensive, reusable, and replaceable. But phototubes are relatively large, thus not many can be packed into a small space, and the image obtained is necessarily lacking in detail. This is illustrated in Figure 10.10, where the numbers represent the photoelectrons actually recorded by each phototube and the dotted lines are the contours of the Cherenkov light image. The detector elements here are 5-centimeter-diameter phototubes with angular diameters in the focal plane of the reflector of 0.5 degree (Figure 10.11). With 37 pixels, the field of view is nearly 3.5 degrees, and the accuracy with which the arrival direction can be measured is +0.2 degree, a tenfold improvement over conventional systems. The limitation in angular resolution is not due to the quality of the optics, but to fluctuations in the electromagnetic cascade that distort the images.

To convert the 10-meter reflector into a large-aperture, fast-exposure, low-resolution camera was not technically difficult. The electronics necessary for the digitizing and recording of optical images were commercially available. However, ground-based, high-energy astrophysics did not have a great attraction for most of the U.S. astronomical community, which usually regarded advances

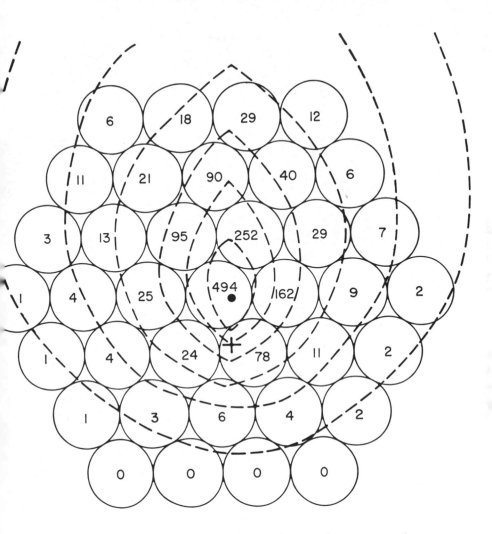

Figure 10.10: *The Cherenkov light image of an air shower as seen by a 37-element camera. The numbers are the light intensities recorded by the individual light detectors; the dotted lines are contours of light intensity.* (Smithsonian Astrophysical Observatory illustration by John Hamwey)

in high-energy astrophysics as synonymous with space research (where the costs are astronomical!). Thus, funding was not immediately forthcoming, and the first practical atmospheric Cherenkov camera was built at the Crimean Astrophysical Observatory

Figure 10.11: *The 37 phototube elements of the 10-meter camera. Each phototube has a diameter of 5 centimeters.* (Smithsonian Astrophysical Observatory photograph)

in 1980. Eventually, an international consortium of experimenters from University College (Dublin), the University of Durham, the University of Hawaii, Iowa State University, and the Smithsonian Astrophysical Observatory obtained the necessary funding. This "10-meter camera" went into operation in October 1983 and has been used routinely ever since.

At the same time, other experimenters both in the United States and overseas are developing technical improvements of both the atmospheric Cherenkov technique and particle array systems. This means that the increased sensitivity to be achieved by orbiting the GRO will be matched by improvements on the ground. Together, these totally different experimental techniques—the sophisticated space satellites, the large optical reflectors, and the distributed arrays of particle detectors—will try to unravel the mystery of the gamma-ray sources: the origin of their prodigious energy output, the way in which the particles are accelerated to these energies, and the mechanism by which the particle energy is transferred to gamma-ray photons.

In contrast to other branches of astronomy, one thing is certain: the gamma-ray universe is a hot, violent place filled with objects that are unstable and short-lived. Rapid variations on all time scales are the norm. In fact, recent reports suggest flaring activity on time scales of minutes from such diverse sources as the Crab pulsar, Cygnus X-3, and Hercules X-1. Some of these outbursts, or flares, involve short, intense emission of trains of photons, separated by some characteristic period; in no case is the flaring activity seen in wavelengths other than high-energy gamma rays. Like the flames that constitute the center of a fire but are only briefly glimpsed through the smoke, these brilliant flares of gamma rays may provide unique clues to the energy processes at the center of these objects. The conditions in these astrophysical furnaces probably are not like anything seen elsewhere in the universe. Indeed, there is a tantalizing possibility that new and potentially important physical processes may be at play. Gamma-ray astronomy is at such a primitive stage of development that we have only a fragmentary picture of what these processes might be.

Although the origin of the cosmic radiation, sometimes considered the "Holy Grail of Astrophysics," is still undiscovered, there is general agreement that the solution may be at hand. For example, Cygnus X-3 is almost certainly a very powerful source of cosmic rays; when we have a better understanding of its power source, we will be able to make an accurate assessment of its cosmic-ray production rate. Some have speculated that the rate is so high that this object alone fills the entire galaxy with cosmic rays. If this proves to be the case, then gamma-ray astronomy can be said to have already made a major contribution to astrophysics.

Humans, the Universe, and Tools for Astronomy

GEORGE B. FIELD

Like all really important ideas, the idea of *universe* is hard to define. By this term, the astronomer means simply, "everything." Of course, for most people—even astronomers—the word "everything" lacks a certain precision. Not only is it philosophically unsatisfying, but it leaves unanswered some serious and tangible questions. For example, are we to count in "everything" that matter that is not observable to us, even in principle? And should we count it in if we believe it will become visible at some time in the future? Most astronomers put such vexing questions aside and, perhaps to their discredit, focus on rather specific *models* of the universe that can be tested directly against observation. Such models also form the core of the field of cosmology, the study of the universe as a whole, including its creation, evolution, and possible end.

Over the years, the various models of the universe have been modified so as to bring them into ever-closer harmony with astronomical observation and the laws of physics. All cosmological models must be based on some generally accepted evidence. Fortunately, astronomical research over several centuries has produced a remarkably detailed historical record for a significant portion of the universe's lifetime.

We know, for example, that the Earth has rocks 4.5 billion years old. (The oldest fossils known are those of algae, 3.5 billion years old. Evidently, life emerged early in the history of the Earth.) Certain meteorites, believed to be shards of one or more planetary bodies that at one time orbited between Mars and Jupiter, also have been dated as 4.5 billion years old, as have the oldest rocks on the Moon. Hence it is usually assumed that all the planets and their satellites formed at that time. According to most theories, the same physical processes that formed the planets would have formed the Sun at about the same time, so the Sun must also be about 4.5 billion years old. The Sun shines by nuclear energy released in its interior, and it can be calculated that there is enough nuclear fuel in the Sun to keep it shining at essentially its present brightness for about 10 billion years.

Astronomers have found approximate ways to date other stars even though they are very distant from us. The range of stellar ages is surprising—from a few million years to about 10 billion years. Evidently, star formation is a continuous process. The youngest stars are often observed to be associated with interstellar gas and dust. Although these materials are found everywhere throughout the Milky Way galaxy, the newborn stars are almost always associated only with the most concentrated clouds of gas and dust, as if parts of such clouds were especially dense and had contracted to form a star.

Because there seems to be a definite upper limit on the ages of stars in the galaxy, astronomers believe that the Milky Way itself must have formed about 15 billion years ago. Some clues about the origin of our galaxy can be gained from study of other galaxies in space. Messier 31, the Andromeda nebula, our nearest large neighbor, appears to be very similar to what we believe our own galaxy would look like could we get far enough away (tens of thousands of parsecs) to view it as a whole (Figure 11.1).

As do stars, galaxies tend to form groups or clusters, held together in a definite region by their mutual gravitational attraction (Figure 11.2). Some groups and clusters of galaxies appear to be stuck together in even larger agglomerations of thousands of gal-

axies called superclusters. But there the tendency to group together ceases. The superclusters are definitely not stuck together, although they seem to be connected by vast sheetlike and stringlike arrangements of galaxies, between which appear to be huge voids containing few if any galaxies. Superclusters are observed to be flying apart from one another at velocities up to half the speed of light. From the distance and recession velocities of superclusters, one can calculate when the superclusters must have been close together. Interestingly, this is about 15 billion years ago, a period of time roughly equal to the age of our galaxy. It is as if a giant explosion 10 to 20 billion years ago flung all matter into space with tremendous force. Soon thereafter, galaxies like our own formed, and because the matter from which they were made had been expanding, the superclusters of galaxies continue to move apart from each other.

Obviously, it is much harder to date whole galaxies than individual stars. But the evidence suggests that however old they are, they all have approximately the same age. This too fits the idea of a massive explosion. When attempts *are* made to get actual ages of galaxies, the estimates again are roughly 10 to 20 billion years ago, and soon after the universe started expanding.

As the clock runs backward and astronomers try to determine events in the first few seconds of creation and expansion, our understanding becomes much more tentative. One of the most popular current cosmological models could be called the "new inflationary Big Bang relativistic model of the expanding universe," or NIBBRMEU for lack of any reasonable acronym.

Why this grandiose name?

Well, the universe *is* "expanding," so that part is obvious.

"Relativistic" refers to the fact that the matter in this model of the universe obeys the principles of the general theory of relativity. This theory, published by Albert Einstein in 1915, is still the most accurate way of describing the large-scale motions of gravitating matter—even when it is moving at a substantial fraction of the speed of light.

"Big Bang" refers to the fact that all the matter in the universe at a particular instant of time was infinitely compressed and, then,

Figure 11.1: *The Andromeda galaxy (Messier 31).* (Lick Observatory photograph)

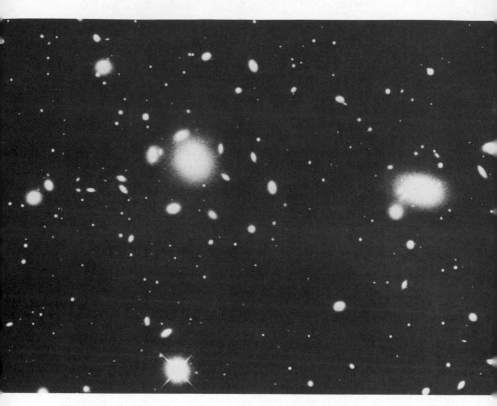

Figure 11.2: *The Coma cluster of galaxies, a system containing over 1,000 members, of which the two supergiant ellipticals are the most evident in this picture.* (Kitt Peak National Observatory photograph)

in an explosion that immediately followed, flung into space. That moment must have occurred 10 to 20 billion years ago, if the model is to agree with observation.

While all the preceding terms, or concepts, have been known for a generation, "inflationary" is a relatively new word in the lexicon of cosmologists. It refers to the fact that contemporary theories of elementary particle physics, or so-called grand unified theories, predict that soon after the Big Bang (only 10^{-35} second later, in fact!), the universe suddenly began to inflate at a high rate, expanding its size by a factor of 10^{25} or more in the next 10^{-35} second. (The word "new" refers to the fact that an earlier theory of the inflationary universe needed modification to ensure agreement with observation.)

Not everyone accepts NIBBRMEU, of course, because there are alternative interpretations of what we see at every step along the way. Roughly speaking, the number of dissenters among knowledgeable scientists increases as you work your way backward from the U to the N in NIBBRMEU. Almost everyone accepts the EU, and most also accept the RM. Only a few cosmologists disagree with the BB. And many cosmologists are still questioning the NI.

Depending on one's viewpoint, then, NIBBRMEU can be characterized as following directly and elegantly from the basic postulates of the general theory of relativity and of the grand unified theory. It predicts the following phenomena:

1. The universe is expanding

2. The universe contains precisely enough matter (both ordinary and exotic) so that it is spatially "flat." This means that the distance between any two galaxies will approach infinity in proportion to the two-thirds power of the time. As a consequence, the universe expands forever but the *rate* of expansion is always getting smaller. The expansion stops for good only when the universe has infinite size. The critical amount of matter needed for a flat universe can be calculated from the currently observed rate of expansion; it is about 10^{-30} times the density of water.

3. The universe contains ordinary matter—neutrons, protons, and electrons, as well as blackbody radiation—but very little antimatter or magnetic charges. (The latter forms of matter could exist in principle, but they were eliminated from the universe by processes early in its expansion.)

4. From the observed intensity of blackbody radiation (equivalent to 3 degrees Kelvin) and from the observed abundance of heavy hydrogen, 75 percent of the neutrons and protons should be in the form of hydrogen, 25 percent should be helium, and minute (but very specific) amounts should be light helium and lithium.

5. The blackbody radiation should be almost uniformly distributed around the sky, with small variations that are calculable.

6. The ordinary matter should be agglomerated into configurations such as clusters and superclusters of galaxies.

All of the predictions, with the exception of 2. have been tested, and none disagrees with observation. Still, the fact that we have not yet found a direct test of the second prediction (2.) bedevils further development of the NIBBRMEU model—and others. Indeed, the observations testing prediction 4. have found that the amount of ordinary matter is only one-tenth the critical amount required for a flat universe. This means that if the model is correct, nine-tenths of the matter in the universe must be invisible!

Five Questions for the Future

Astronomers today continue to search for a model that will accurately describe the universe's beginning—and predict its possible end. Indeed, although there is a vast array of research problems of concern to astronomers, most ultimately lead back to these basic questions. To underscore this point, I have chosen five specific problems that are fascinating to me personally, but that also promise some likely solutions in the next 10 years. They are drawn from the report *Astronomy and Astrophysics for the 1980s* published by the National Academy of Sciences in 1982. This was the result of a two-year effort by a committee of astronomers I chaired to identify future opportunities for astronomical research and to recommend the facilities needed to carry them out.

The five questions I would like to consider further are these: Are there more than nine planets in the universe? Are stars and planets still forming today? Did all galaxies form soon after the Big Bang? Is there a substantial amount of hidden matter in the universe? And, what is the nature of this matter? In each case, new facilities and instrumentation—*new tools for astronomy*—will be required to provide the evidence necessary to answer the questions.

Since this book has been concerned with the new tools designed to address these and other specific questions, my brief discussion may reinforce the sense of inquiry and questioning that underlies the demand for new instrumentation. At the same time, however,

I hope to demonstrate again how modern scientific discovery and understanding have become directly linked to technological advances.

Of course, we know there are at least nine planets—we observe eight in the Solar System, and we are standing on the ninth. If there *is* a tenth planet in the Solar System, it must be very small. The skies have been searched carefully since the last planet, Pluto, was discovered in 1930, and nothing has been seen. In other words, if other planets exist, they must be in orbits around stars other than the Sun.

Finding planets near other stars is dreadfully difficult because planets are so much fainter than their parent stars. Even Jupiter, our largest planet, is only a billionth as bright as the Sun. To detect even the isolated image of such a faint object many light-years from the Earth would be very hard with current techniques; but it is nearly impossible to photograph such a planet in the presence of its parent star. The light of the star simply spills over into the area where the planetary image would be expected. A better technique is to look for the displacement in the position of the parent star caused by the gravitational pull of the orbiting planet. The predicted effect for a Jupiter-like planet orbiting a Sun-like star that is 10 parsecs from the Sun is a periodic motion of about 10^{-3} arcsecond. (An arcsecond is $\frac{1}{3600}$ of a degree, equivalent to the thickness of a piece of paper seen at a distance of 10 meters.) Amazingly enough, current ground-based telescopes can achieve an accuracy of 3 milliarcseconds in measuring these motions, and improvements in instruments promise to reach 1 milliarcsecond within the decade. Even beyond that, large improvements should be possible with instruments placed in Earth orbit, where the lack of atmospheric blurring will yield much sharper images of stars. The Hubble Space Telescope, with images 20 times sharper than those made by telescopes on the ground, will make a first step in this direction when it is launched later in this decade. COSMIC, an optical interferometer for space of the type recommended for the 1990s by the National Academy and described earlier by Wesley Traub (page 79), should be capable of resolving 0.03 milliarcsecond.

As we look for other planets, we may answer the related question: are new planets and planetary systems like our Solar System

forming at the present time? Recently, the Infrared Astronomy
Satellite (IRAS), a joint project of the United Kingdom, the Netherlands, and the United States, surveyed the entire sky at far-
infrared wavelengths between 8 and 120 microns. Among the results were magnificent images of celestial objects and phenomena
normally invisible from the Earth.

Infrared radiation is emitted by materials that are warm but not
hot, such as dust in interstellar space and in orbit around stars (see
Willner, page 154). IRAS images demonstrated conclusively what
had already been learned by fragmentary earlier observations: the
interstellar dust in the Milky Way forms a thin layer across the
plane of the galaxy, where it is heated by starlight to glow in the
infrared (Figure 11.3). Scattered throughout this layer are hot spots
that appear to be regions containing dense interstellar matter,

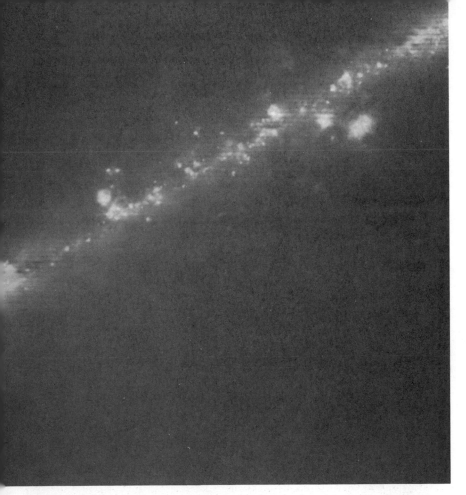

Figure 11.3: *This image of the center of our galaxy was produced from observations made by the Infrared Astronomical Satellite (IRAS). The bulge in the band is the galactic center. The knots and blobs scattered along the band are giant clouds of interstellar gas and dust heated by nearby stars. Some are warmed by newly formed stars in the surrounding clouds, and some are heated by nearby massive, hot, blue stars tens of thousands of times brighter than the Sun.* (Jet Propulsion Laboratory photograph)

including dust, heated by stars that are extraordinarily luminous (100,000 times as bright as the Sun). However, because the dust in the Milky Way is so opaque, we can't see the light from these stars at ordinary visible wavelengths. All we see is the infrared radiation from the nearby dust. Here are stars that have formed out of the surrounding interstellar matter within the last few million years. They must be this young because stars so luminous cannot last much longer, given their limited supply of nuclear fuel.

Both ground-based infrared and radio astronomers have studied one of these regions—the Kleinmann-Low nebula in Orion—for some time. The pattern of emission and of velocities in the region strongly suggests that one of the objects is less than 10,000 years old. The evidence is based upon high-velocity motions indicating the ejection of matter from the object. Because the motions have not yet carried the matter very far, one infers that the object is still very young (by usual astronomical standards). In short, stars must have formed in the Orion region in the relatively recent past.

What about planets? Although our techniques are still crude, there is increasing evidence provided by some exciting ground-based observations that planets may be there, too. For example, the flows of gas associated with newly formed stars often show a highly asymmetric pattern, with gas flowing outward in opposite directions along a single straight line through the new star. Why should the gas flows be confined to a line? One interpretation is that a disk of gas and dust is in orbit around the star, thus forming a disk of matter similar to Saturn's rings. The young star, being very luminous, repels gas in its vicinity by radiation pressure, but the material in the disk is so dense that it resists the impact of the outward-streaming gas. The gas simply streams out perpendicular to the disk, where the resistance is least, thus explaining the fact that the observed expansion is along a line. Interestingly, a disk of gas and dust has long been hypothesized as the original birthplace of the planets in our own Solar System. Perhaps, then, we are seeing the potential birthplaces of new planetary systems around these young stars.

As confirmation of this argument, IRAS also discovered that there is a cloud of small particles in orbit around the bright star Vega, as well as similar clouds around other stars. Although Vega is quite an old star (100 million years or so), it appears to have retained a remnant of the type of disk discussed above. Although the particles in the Vega cloud must be very small, they could coalesce to form larger bodies, such as planets; in fact, we don't understand why they have not already done so.

Following up the IRAS discoveries, astronomers Bradford Smith and Richard Terrile in October 1984 reported ground-based observations of planetary-type material around the infrared star Beta Pictoris (Figure 11.4). This Southern Hemisphere star, about two times as massive as the Sun, was identified by IRAS as probably surrounded by dust and gas. Using electronic enhancement techniques, Smith and Terrile confirmed the existence of the material and that it is concentrated in a disk.

Then, just two months later, Donald McCarthy and Frank Low of the University of Arizona and Ronald Probst of Kitt Peak National Observatory reported making direct observations of a planetlike object orbiting a dim star known as Van Biesbroeck 8. The object, slightly smaller than Jupiter but much more dense, does not seem to be a planet in the sense of those bodies orbiting the Earth. Rather, it may be a "brown dwarf," that is, a small star lacking sufficient mass to trigger nuclear fusion processes at its core. Such objects had been predicted by theory, and several possible examples had been identified earlier, but this observation is the first clear-cut evidence for their existence. The parent star, Van Biesbroeck 8, had been selected as a likely candidate for having a planetary system because of minute displacements observed in its motion. The discovery, however, came only after a yearlong, painstaking search with special instruments.

What are the prospects for future research in this area? Infrared astronomers have proposed a much more powerful facility to follow up the IRAS survey. As described in the chapter by Steven Willner, the Space Infrared Telescope Facility (SIRTF) will be a 1-meter-diameter telescope cooled, as was IRAS, to liquid-helium temperatures. Unlike IRAS, which simply scanned the sky, SIRTF can be pointed at any desired target. Because of this and its large aperture, SIRTF will be about 1,000 times more sensitive than IRAS, enough more to permit radiation to be split into various wavelengths, producing spectra. For astronomers, the distinctive pattern of spectral lines produced by each atom or molecule serves as the fingerprint of a celestial object, carrying clues to its physical and chemical nature. Because the regions under investigation are thought to contain a large variety of molecules, the spectral information may reveal the chemical composition of the emitting regions. Equally important, the relative strengths of the lines can be analyzed to determine the temperature and density of each region. And the Doppler shifts of the lines will also permit astronomers to determine the direction and velocity of the matter in these regions. It will be particularly interesting to see whether the emitting material forms a disk around one of the young stars, as predicted by theorists interested in the formation of planets.

As spectacular as SIRTF will be, it can be improved upon with respect to its spatial resolution. It is important to obtain sharp images of emitting objects; otherwise, we cannot be sure we are studying a single large object or a group of many small objects. Producing sharp images is harder at infrared wavelengths than in the optical domain because the angular resolving power of a telescope is inversely proportional to the wavelength. The only solution to this dilemma is to build infrared telescopes with larger apertures. Because the Earth's atmosphere is opaque to most infrared wavelengths, this requires that the large infrared telescope be operated in space.

Putting a very-large-aperture telescope in orbit is a formidable technical challenge because the cargo bay of the Space Shuttle is not wide enough to carry directly into orbit instruments much larger than the Hubble Space Telescope or SIRTF. Instead, various small modules must be put into orbit and later assembled by astronauts. NASA is eager to attempt this and has formulated a plan for what they call the Large Deployable Reflector (LDR), a giant infrared telescope with an aperture of 10 meters or more. (See Ho, page 148.) Such an instrument would have an angular resolution of better than 2 arcseconds when operated at a wavelength of 100 microns. If the telescope operated at even shorter infrared wavelengths, the angular resolution would be better still. While not

Figure 11.4: *What may be another Solar System 50 light-years away is seen in this picture taken by astronomers at the University of Arizona and the Jet Propulsion Laboratory. The charge-coupled device (CCD) picture of Beta Pictoris shows a circumstellar disk of material extending 40 billion miles from the star, which is located behind a special occulting mask in the center of the photo. (Dark lines are filaments supporting the mask.) The material is probably composed of ices, carbonaceous organic substances, and silicates. The disk is believed to be fairly young, possibly no more than a few hundred million years old.* (Photograph from the University of Arizona and the Jet Propulsion Laboratory)

spectacular by the standard of the Hubble Space Telescope (0.05 arcsecond), the angular resolution of LDR will nevertheless permit us to routinely observe structures as small as 400 astronomical units across in the nearest star-forming regions. This is about the size of the suspected planetary disk seen by Smith and Terrile. (For comparison, our Solar System is about 90 astronomical units in diameter.) If we are lucky, we may find many more disks of gas and dust orbiting stars, as well as other examples of planets.

Although radio wavelengths are much longer than infrared wavelengths (for example, 1 centimeter is 100 times 100 microns), radio astronomers have achieved extremely high angular resolution by applying the technique of interferometry, that is, increasing their effective aperture by using two or more telescopes separated by some distance to observe the same object. The principles and observing techniques involved have been described in more detail by Wesley Traub and Mark Reid.(See Chap. 4 and 5.) Here I refer only to the basic formula, that an angular resolution of x arcseconds can be achieved at wavelengths of y centimeters by using two telescopes separated at a distance $(d) = 200,000\ y/x$ centimeters. (The same principle applies to individual telescopes; thus, to achieve an angular resolution of 2 arcseconds at a wavelength of 100 microns [0.01 centimeter], the diameter of the telescope must be 1,000 centimeters [10 meters]; this is just the size proposed for LDR.)

Radio astronomers also have shown it is possible to carry out interferometric studies even if there is no physical connection between the two telescopes, if one records the output of each telescope and compares the data later in a computer. The greatest possible separation on the surface of the Earth is just its diameter, 13,000 kilometers, so our formula implies that at a wavelength of 1.3 centimeters (which is that of a powerful emission line of H_2O molecules in space), one can in principle obtain an angular resolution of 2×10^{-4} arcsecond—vastly better than by any other technique. In fact, radio astronomers have already approached this theoretical maximum in several difficult and time-consuming experiments.

Recognizing the power of such radio interferometry, the National Academy gave high priority to the construction of an array of ten 25-meter-diameter radio telescopes to be erected at various sites across the United States, including Puerto Rico and Hawaii.

This Very Long Baseline Array (VLBA), described earlier by Reid (page 120), will permit astronomers to obtain images of radio-emitting objects with routine precision of 0.002 arcsecond. Applied to the problem of star formation, the VLBA will enable us to resolve details only 0.05 astronomical unit across in the nearest star-forming regions. It happens that such regions emit the H_2O line at 1.3 centimeters very powerfully, so it should be possible to build up a picture of what is happening in such regions. Perhaps one or more cases can be found to test the theoretical prediction that a disk of gas and dust is a precursor to planet formation.

This brings us to our third topic, galaxy formation. All evidence suggests that galaxies like our own (in which stars are still forming) were themselves formed in the remote past, perhaps soon after the original Big Bang explosion. Astronomers studying galaxies at great distances recently discovered a galaxy with a velocity of recession 60 percent that of the speed of light. Quasars, which are believed to be exploding galaxies, have been observed with velocities up to 90 percent that of light. According to the models of the expanding universe, the light from the distant galaxy began its journey toward us when the universe was only one-half its present size and only one-third its present age; the corresponding figures for the most distant quasars are one-fifth and one-tenth, respectively. Thus, by peering out into space, astronomers are also looking backward in time, when the universe was not only younger but smaller.

In principle, we should be able to observe regions of the universe so distant that the galaxies were just forming when the light left them. Various theories indicate that this galaxy-forming epoch occurred when the universe was roughly 15 percent its present size and 5 percent its current age—perhaps just beyond the region where the most distant quasars are now seen. In fact, it is not implausible that the quasar phase occurs early in the life of galaxies. If this view of galaxy formation is correct, we would expect to observe an upper limit on the recession velocities of quasars, that is, a region of space beyond which no quasars are seen. In fact, such an upper limit is observed and it fits tolerably with the theory of galaxy formation. We just don't yet know if this is its explanation.

How do we test theories of galaxy formation? The answer very much depends upon the wavelength of observation. The Hubble

Space Telescope, for example, should be able to study routinely galaxies with velocities up to 60 percent of the speed of light. Moreover, it may be able to look deep enough to see other objects receding at 80 percent of the speed of light, or an age when the universe was one-third its current size. Objects more distant, however, emit radiation outside the bandwidth to which the Hubble Space Telescope is sensitive, so other approaches seem more likely to succeed. In particular, SIRTF will be sensitive to infrared emission, and it could possibly detect the visible light Doppler-shifted into the infrared band as it is emitted from galaxies in the process of formation.

Another possibility is offered by the proposed Advanced X-ray Astrophysics Facility (AXAF), a 1.2-meter X-ray telescope to be launched by the Space Shuttle. If funded by the federal government, this facility, given the very highest priority by the National Academy, could be flown in the early 1990s. Here is how AXAF can help. Jeremiah Ostriker of Princeton University and his collaborators have calculated that when a galaxy is forming, there should be a tremendous number of stellar explosions of the type referred to as supernovas (Figure 11.5). Supernovas occur when the cores of rapidly evolving, recently formed, massive stars collapse to form a neutron star or a black hole. The collapse is accompanied by the release of huge amounts of energy. Three such explosions—in A.D. 1054, 1572, and 1604—have been observed in our own galaxy. Dozens more have been inferred by the supernova shock waves still seen expanding into the interstellar medium.

In our own galaxy, the rate of supernovas is low enough that the energy of each explosion is readily dissipated by the surrounding medium. This is because there is relatively little interstellar material in our galaxy, most of it having been used up by 10 billion years of star formation. But in a young galaxy, virtually all the matter may still be interstellar, so the rate of star formation is correspondingly large, perhaps 100 times greater than in the Milky Way. Under these conditions, the rate of supernova explosions, which follow a few million years behind the formation of massive stars, will be 100 times the present rate, or several supernovas per year. The shock waves from these multiple explosions will overlap before the energy can dissipate, and the whole remaining interstellar medium will be heated to an incredible 10^9 degrees. The normal gravitational force of a galaxy is unable to retain such a

Figure 11.5: *The Crab nebula, a supernova observed by Chinese astronomers in 1054 and the subject of intense study by modern astrophysicists.*
(Harvard College Observatory photograph)

hot gas, so the gas will stream out of the galaxy at thousands of kilometers per second. Such a gas is a copious emitter of X rays, and AXAF ought to be able to detect it.

The Einstein Observatory, which operated between 1978 and 1981, imaged a number of very distant quasars with no difficulty. AXAF will be 50 times more sensitive than Einstein, and so should be able to image the galaxies predicted by Ostriker at somewhat greater distances still.

Why do we care about galaxy formation? Well, beyond the intrinsic interest in understanding the origin of our own galaxy, there is the complex issue of the behavior of gravitating matter in the early universe. And this brings me to the two last questions: invisible matter in the universe and its nature. According to NIB-BRMEU, 90 percent of the matter is hidden from view, and hence presumably in some exotic form other than protons, neutrons, and electrons. In that case, the evolution of gravitating matter is dominated by exotic matter, with ordinary matter going along for the ride.

Actually, the concept that there is hidden matter in the universe is really not new. In the 1930s, the astronomer Fritz Zwicky suggested "invisible matter" as the solution to a puzzling discrepancy reported by astronomers observing the positions and Doppler motions of galaxies in a cluster of galaxies. Calculations of the amount of matter required to hold the cluster together by gravitation consistently produced figures larger than the amount of matter visible in the system. Over the years, alternative forms of "ordinary" matter—for example, ionized gas between the galaxies—have been sought to explain the discrepancy, all with no luck. Most recently, optical studies of spiral galaxies and X-ray observations of elliptical galaxies suggest the presence of outer envelopes of invisible matter with about 10 times the mass of the galaxies' luminous disk. Some of this must be ordinary matter (protons, neutrons, and electrons), if we judge by the nucleosynthesis data on the early universe. What form this ordinary matter might take is uncertain; maybe it is in very low-mass stars that are too faint to observe. Perhaps SIRTF can observe the infrared light of these stars, because they are relatively cool. (Alternatively, AXAF might observe their X-ray emission. The Sun's corona emits copious X rays as a consequence of its high temperature [1,000,000 degrees Celsius], and, ironically, data collected by the Einstein Observatory indicate that this effect may become more pronounced for "cool" stars of even lower mass.)

Alas, even this invisible matter on the outskirts of galaxies is not enough to fulfill the prediction of the NIBBRMEU model; another 10 times more matter must exist in the form of exotic particles. Such particles might even reveal their presence directly, for example, by decaying into observable protons. On the other hand, we must be prepared for the eventuality that these exotic particles will simply remain unobservable. Yet their presence is an important prediction of physics.

What to do? For now, the only alternative is to search for the gravitational effects of the hidden mass. In clusters of galaxies, for example, the observed gravitational phenomena could be caused by large numbers of faint stars made of ordinary matter, such as the brown dwarf found orbiting Van Biesbroeck 8. But similar observations of superclusters of galaxies suggest there is more hidden matter than can be accounted for simply by dim stars. While the required amount is twice the 10 percent known to be present as ordinary matter, it is still about five times short of the critical amount. The final question, then, is, what is the nature of this invisible matter?

An early speculation suggested the "exotic matter" might be neutrinos, a charged particle usually assumed to have a rest mass equal to zero. However, some Soviet experiments on beta-decay have suggested that the mass of the neutrino is about 1/20,000 of that of the electron. Although this mass is tiny in an absolute sense, it is enough, given the vast numbers of neutrinos predicted to exist in the Big Bang model of the universe, to contribute a substantial fraction of the critical density. It is possible to compute the fate of a universe made largely of massive neutrinos. Such a theoretical universe actually simulates certain aspects of the observed universe, particularly in making large sheets containing superclusters of galaxies with large voids in between (Figure 11.6). However, the model fails to agree with other aspects of the universe, such as the fact that the galaxies apparently formed long ago and the superclusters are still forming. As a consequence, massive neutrinos don't seem like the answer; maybe the Soviet experiments are incorrect.

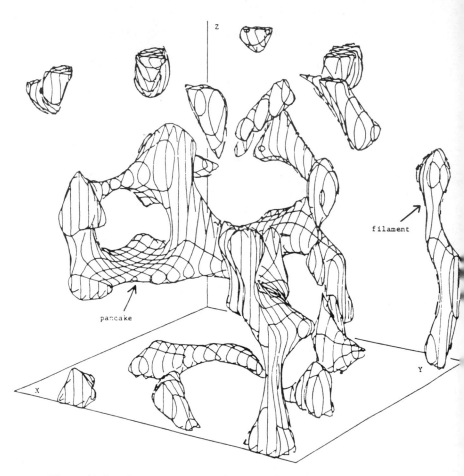

Figure 11.6: *A computer-created numerical representation of the neutrino universe.* (Illustration courtesy of Joan Centrella; and Adrian Melott; reprinted by permission from *Nature*, Vol. 305, p. 196, Copyright 1983.)

To form galaxies, it is important that the basic particles have very low random velocities. There are several candidates that meet this requirement, among them the photino, a particle rather like a photon of light but with different spin properties. The photino is consistent with the grand unified theory; in fact, such a particle is required if one adds to that theory the requirement that it be symmetric under transformations from particles of half-integral spin to full-integral spin and vice versa, a principle called *super-*

symmetry. (Recent experiments at the European Center for Nuclear Research have revealed some fascinating events involving high-energy particles which theorists believe may be explained by the predicted photino.)

Suppose photinos existed and comprised 90 percent of the mass of the universe. How would we know it? Well, particle physicists think they can predict the magnitude of fluctuations in the density of the universe on various scales. It is then relatively straightforward to calculate how these fluctuations would be amplified by gravitation as the universe expands. The first test comes at the point where the universe was one-thousandth of its present size and 0.00003 of its present age. Such an epoch is far beyond the reach of our largest telescopes, but according to the theory, it should leave its imprint upon the distribution of the blackbody radiation over the entire sky. Indeed, radio astronomers already have found a large-scale (180-degree) variation of low amplitude across the sky. However, this variation is probably due entirely to the peculiar velocity of our galaxy and the resulting Doppler shifts. No other variations have been seen on any other scale, down to small fractions of a degree, but because the predicted variations are so small, the results are still in accord with prediction.

The definitive test of variations in the blackbody radiation should be provided by the Cosmic Background Explorer (COBE) satellite to be launched by NASA in 1987. This dedicated (with a single purpose) mission will determine variations on angular scales down to 7 degrees or so. In addition, the LDR mission will determine variations on smaller scales corresponding to those of galaxies and clusters of galaxies.

Unfortunately, the photino hypothesis does not provide a ready explanation for the observed fact that superclusters tend to occur in sheets, leaving vast voids between them (Figure 11.7). In the photino hypothesis, small objects form first. Stars and star clusters appear and then later agglomerate into galaxies and clusters of galaxies, so there is no obvious way *to prevent* galaxies from forming. Moreover, because photinos are the first things to contract, providing the centers around which galaxies form, observers should find 100 percent of the critical density of matter when they examine gravitational phenomena on the scale of superclusters. So far, they have not done so.

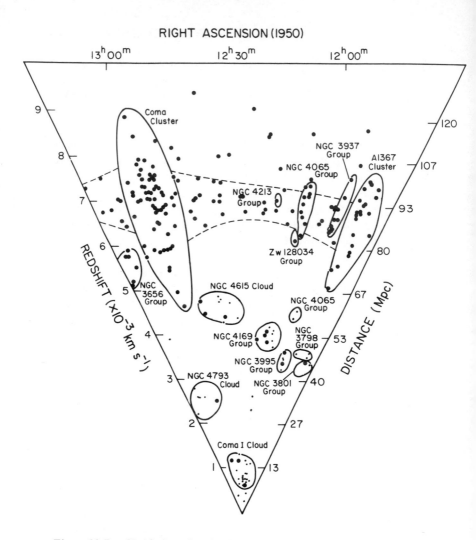

Figure 11.7: *Voids found in the distribution of clusters of galaxies through a systematic survey of galactic redshifts.* (Harvard-Smithsonian Center for Astrophysics illustration)

Along with David Spergel and Alan Guth at the Harvard-Smithsonian Center for Astrophysics, I have been exploring the implications of supernova explosions for the photino model. Following a suggestion of Nick Kayser, we suppose that the collapse of photinos into clumps to form stars and galaxies occurs only when there is first a favorable large-scale (supercluster-sized) positive fluctuation in the matter density. This effect will lead to the formation of some galaxies and the accompanying high supernova rate predicted by Ostriker, but in this model, the explosions should be concentrated in the region of large-scale positive density fluctuation. The supernova-heated gas will stream out of the parent galaxies and merge to form huge bubbles of hot gas, which, in turn, will expand into the surrounding regions, including those of both positive and negative large-scale density fluctuations. Encounters with positive density fluctuations will have relatively little effect; they are already boiling over with supernova explosions anyway. But heating the ordinary gas in negative density fluctuation regions could have the effect of preventing galaxy formation there. To fall into the gravitational-potential wells formed by photino clumps, gas must be cooler than about 10,000 degrees Celsius; thus, heating the gas to temperatures between 100 million and 10 billion degrees Celsius would effectively halt galaxy formation. The result would be large-scale regions filled with galaxies, intermingled with other regions in which few, if any, visible galaxies exist. These dark voids would still be filled with large numbers of photinos; in fact, some 80 percent of all the photinos would be here, clumped into stellar and galactic-sized aggregations but lacking any of "ordinary" matter necessary to render them visible. In other words, such a universe would be composed largely of "shadow galaxies" lurking in the voids between superclusters.

Naturally, the preceding model depends upon very complex and detailed astrophysical processes. Our model is still early in its formulation, and chances are it will not survive intact. But I am intrigued by one consequence: the X-ray emission predicted to accompany such a heating process. For nearly 25 years, it has been known that a diffuse X-ray background permeates the entire sky. The Einstein Observatory's data indicated that this background radiation may be produced by millions of so-far unresolved sources, but most likely distant quasars. On the other hand, an earlier X-ray satellite measured the spectrum of this background and found it fitted well with radiation of a gas whose temperature is 500 million degrees Celsius. Could this be the supernova-heated gas streaming out of galaxies as predicted by Ostriker? And could that gas quench further galaxy formation? AXAF, the highest-priority new space instrument for the next decade, will help us decide by observing the sources of the diffuse X-ray background and determining their spectra.

Humans in the Universe

Most astronomers I know *think* that we are not alone, or, in other words, that humans are not the only intelligent species in the universe that is capable of communicating. But no one *knows* whether we are alone, because the search for extraterrestrial intelligence has only just begun.

Recently, however, there has been a major improvement in the rate at which searches for radio signals of intelligent origin can be carried out. Paul Horowitz of Harvard University has developed a receiver with 10,000 independent channels, which, when attached to an astronomical radio telescope, can look for signals over a wide-frequency band. The National Academy recommended an increase of support in this area, and the National Aeronautics and Space Administration has complied by starting to develop a receiver much more efficient still. If a tenth planet were discovered, for example, there is no doubt that radio astronomers would point their largest instruments at it and tune in their most efficient receivers in hopes of discovering signs of life.

We must admit that the chances that a Jupiter-like planet would harbor intelligent life may not be very good. After all, to be detectable with present techniques, a planet must be massive, and the two massive planets in our own system, Jupiter and Saturn, show no signs of intelligent activity. However, the search for extraterrestrial intelligence (SETI, for short) need not be limited to known planets. The beams of radio telescopes are so large that they cover many potential planetary sites at once. Thus, the search can be done quite differently: use radio telescopes to seek out intelligent signals; then if such signals are detected, study those stars in the radio beam for the effects of planetary gravitational forces.

Of course, detection of extraterrestrial intelligence would have profound implications for the whole of humanity. But astronomers

Figure 11.8: *The giant meteor crater near Winslow, Arizona.* (Smithsonian Institution photograph)

have some basic questions they would address should such signals be detected. Detection alone could yield an orbital period (from the Doppler shift of the signals) and hence a distance from the parent star to the planet. Details of the Doppler shift would also allow one to derive the eccentricity of the planet's orbit. Moreover, the signals themselves might yield information on other physical properties of the planet, as well as the chemical basis for life there. Such questions are of fundamental importance in discovering how widespread life may be in the galaxy.

Oddly enough, astronomical research may also provide clues to how life developed and evolved on this planet. For example, a recent series of discoveries seem to challenge the long-held notion of biological evolution as a series of imperceptibly small steps. Paleontologists have found long-term (approximately 26 million years), quasi-periodic behavior both in the rate of extinction of species—as revealed by the fossil record—and in the rate of formation of large, nonvolcanic craters on the Earth—as revealed by radioactive dating of the rocks in their vicinity (Figure 11.8). Not only do the periods involved in the two phenomena appear to be of the same order of magnitude, but the two phenomena appear to be roughly synchronized, with the rate of extinction increasing whenever the rate of cratering does.

An unusual cosmic connection between these phenomena was found in 1975, when Luis and Walter Alvarez showed that the thin layer of geological strata corresponding to the boundary between the Cretaceous and Tertiary eras—a period 65 million years ago marked by massive extinctions, including that of the dinosaurs—contains an extremely high abundance of iridium, a chemical element that is otherwise very rare on Earth. But iridium is found in meteorites, masses of rock and iron that are thought to be the debris of planetary collisions or, in some cases, primeval material left over from the formation of the Solar System. Such material falls regularly on Earth from space, sometimes (very infrequently) with a mass sufficient to cause cratering of the Earth's surface. Based on the Alvarezes' finding, some scientists have hypothesized

Figure 11.9: Comet West (1976), with its irregular ion tail (above) and smooth dust tails (below). (Photograph courtesy of Jack W. Harvey, Kitt Peak National Observatory)

that 65 million years ago, one or more giant meteorites (or perhaps the nuclei of comets, which have a similar composition to that of meteorites) impacted on the Earth. Too massive to be slowed down by the atmosphere (as are ordinary meteorites), it plunged into the ground at speeds up to 20 miles per second, sending shock waves into the rock, pulverizing it, and ejecting great dust clouds into the atmosphere. According to this scenario, the dust was so thick that sunlight was cut off, thus preventing photosynthesis and killing many of the plants that formed the base of the food chain. The result was widespread starvation for many species and, for some, extinction. There is much evidence that large craters now seen on the Earth's surface were actually caused by the impact of massive bodies. Calculations of the amount of dust created by such impacts, and of its suspension in the atmosphere, support the idea that mass extinctions could have resulted. Although one should maintain a healthy skepticism, we thus have a plausible explanation for the extinction of the dinosaurs, as well as other species seen with a periodicity of about 26 million years.

As in so many branches of science, even a tentative hypothesis raises almost as many questions as it answers. In this case, one must ask what could cause the rate of infall of meteorites to increase every 26 million years or so. Two suggestions have been advanced, both of which require the "meteorite" to be the nucleus of a comet, a few kilometers in diameter (Figure 11.9). There are believed to be billions of cometary nuclei in the outer Solar System, but only a few are perturbed enough by passing stars to enter the inner Solar System, where the Sun's heat and light render them visible. However, if there was some slowly varying gravitational force, it might cause a periodic increase and decrease in the number of comets entering the inner Solar System, and hence, as required by the hypothesis, the number colliding with the Earth.

What could be the source of such a gravitational force? One suggestion is that the Sun, like a large fraction of the stars we observe, is actually part of a binary system. The Sun's companion star—if it exists—might be more than 1.5 light-years distant on a 26-million-year orbit. If this star, dubbed *Nemesis* by the originators of this idea, is of low mass, it would now be too far away for its image to be distinguished from those of millions of other stars of comparable apparent brightness. However, if Nemesis has the right period and moves in an eccentric orbit around the Sun, it too could perturb the comets in the right way. A concerted search for Nemesis is now underway.

Perhaps evolution is punctuated by periods of severe environmental stress, which turn up the pressure on natural selection. The less-adaptable species fall by the wayside, and a new period of biological evolution, characterized by a new set of dominant species, begins. Could that source of stress be dust kicked up by colliding comets, and the rate of such collisions governed by the periodic reappearance of a stellar companion to the Sun? This whole problem is still too little understood to say "yes" to these questions, but, at least, the oft-used phrase "Humanity and the Universe" seems to have acquired new meaning. Perhaps all species, humans included, have evolved largely because of influences from outer space. The new tools for astronomy described in this book may provide the clues to understanding both the cosmos—and ourselves.

Contributors

John Carr is head of the Charles Hayden Planetarium at Boston's Museum of Science. With a background in both science and education, he is concerned with creating links between research institutions and museums, and was instrumental in establishing a Planetarium Advisory Committee in Boston.

James Cornell is Publications Manager of the Harvard-Smithsonian Center for Astrophysics, where he is responsible for programs of technical and public information. He is the author or editor of more than a dozen books for general readers, including *The First Stargazers* (Scribners) and *Astronomy from Space* (MIT Press).

George B. Field is Senior Scientist at the Smithsonian Astrophysical Observatory and the Willson Professor of Astronomy at Harvard University. He was founding director of the Harvard-Smithsonian Center for Astrophysics and Chairman of the Astronomy Survey Committee of the National Academy of Sciences. He is coauthor of *The Invisible Universe* (Birkhäuser).

Paul T. P. Ho is an Assistant Professor of Astronomy at Harvard University. In addition to submillimeter-wave astronomy, his research interests include the interstellar medium and the study of energy sources at the cores of active galaxies. In 1984, he was named the recipient of a Sloan Research Fellowship.

David W. Latham is Associate Director for Optical and Infrared Astronomy at the Harvard-Smithsonian Center for Astrophysics and a lecturer on astronomy at Harvard University. His research includes spectroscopy and photometry of stars, galaxies, and quasars. He is internationally recognized for his development of astronomical instrumentation and low-light detectors.

Alan P. Lightman is a lecturer in astronomy and physics at Harvard and a physicist at the Harvard-Smithsonian Center for Astrophysics, and is primarily concerned with the theories of the physics and dynamics of compact objects, including quasars and black holes. Lightman also writes widely on science and society, including a monthly column in *Science 85* magazine. He is the author of *Time Travel and Papa Joe's Pipe* (Scribners) and coeditor of *Revealing the Universe* (MIT Press).

Mark J. Reid is a radio astronomer at the Harvard-Smithsonian Center for Astrophysics. His research includes the investigation of molecular clouds, circumstellar dust shells, newly forming stars and primitive stellar nebulas, and radio galaxies and quasars using techniques such as very long baseline interferometry.

Robert Stachnik is a physicist and research associate at the Harvard-Smithsonian Center for Astrophysics, with a particular interest in techniques for improving astronomical images. He worked with Antoine Labeyrie on the development of speckle interferometry, a procedure for achieving extremely high spatial resolution from ground-based optical telescopes. This work has expanded to include the study of optical interferometers, using multiple spacecraft.

Wesley A. Traub is a physicist at the Harvard-Smithsonian Center for Astrophysics who has conducted experiments in optical spectroscopy of the Sun, planets, stars, and interstellar medium, as well as far-infrared spectroscopy of the Earth's ozone layer, using both ground-based and balloon-borne telescopes. He is now engaged in building a ground-based Imaging Stellar Interferometer and in the design of an orbiting array of optical telescopes known as COSMIC.

Trevor C. Weekes is an astrophysicist at the Smithsonian's Fred Lawrence Whipple Observatory on Mt. Hopkins, Arizona, where he established and for many years has operated one of the world's most sensitive ground-based gamma-ray detectors. Weekes is the author of *High-Energy Astrophysics* (Chapman and Hall).

Steven P. Willner is an astronomer at the Harvard-Smithsonian Center for Astrophysics, where he pursues research in infrared astronomy, with a special interest in compact sources such as the galactic center and star-forming regions of molecular clouds. He is currently a member of the team from Smithsonian and four other organizations developing an instrument for SIRTF, the Space Infrared Telescope Facility.

George L. Withbroe is an astrophysicist at the Harvard-Smithsonian Center for Astrophysics whose research involves solar physics, with an emphasis on ultraviolet wavelengths. He has developed models for describing the photosphere and corona of the Sun. He is currently developing observation programs for both ground-based and satellite-borne solar instruments.

Martin V. Zombeck is a physicist at the Harvard-Smithsonian Center for Astrophysics, where he was project scientist for the Advanced X-ray Astrophysics Facility (AXAF). He is now the Smithsonian project scientist for the high-resolution X-ray imager on the German Roentgen Satellite (ROSAT), which is to be launched in 1987. He is the author of *The Handbook of Space Astronomy and Astrophysics* (Cambridge University Press).

Further Reading

Because most of the instrumentation and techniques described in this book represent proposed or state-of-the-art advances, a traditional bibliography is not entirely appropriate. The authors have drawn their descriptions of new tools primarily from blueprints, design sketches, research proposals, and their own ongoing work. What follows, then, is a list of selected books and articles that provide additional background in astronomy and astrophysics, some insight on the relationship between technology and science, and further glimpses into the future of research and discovery.

General Astronomy and Astrophysics

Barrow, John D., and Joseph Silk. *The Left Hand of Creation*. New York: Basic Books, 1983. (An overview of the Big Bang and expanding universe cosmology.)

Berman, Louis, and J. C. Evans. *Exploring the Cosmos*. Boston: Little Brown, 1983. (Introductory college text, with class problems and discussions, but will serve as an excellent reference for general readers.)

Frazier, Kendrick. *Our Turbulent Sun*. Englewood Cliffs, N.J.: Prentice-Hall, 1982. (Solid journalistic overview of solar physics, from astronomy to energy.)

Friedman, Herbert. *The Amazing Universe*. Washington, D.C.: National Geographic Society, 1975. (Slightly dated but still good general introduction to the new view of the universe by a pioneer in space science.)

Guth, Alan H., and Paul J. Steinhardt. "The Inflationary Universe," *Scientific American*, Vol. 250, May 1984.

Henbest, Nigel, and Michael Marten. *The New Astronomy*. New York: Cambridge University Press, 1984. (Stunning color reproductions throughout.)

Kaufmann, William J. III. *Universe*. New York: W. H. Freeman, 1984. (Introductory text, unusually well illustrated.)

Spitzer, Lyman. *Searching Between the Stars*. New Haven, Conn.: Yale University Press, 1982. (A discussion, sometimes technical, of interstellar material.)

Trefil, James. *The Moment of Creation*. New York: Scribners, 1983. (Lucid and popular approach to current theories of cosmology.)

Whipple, Fred L. *Orbiting The Sun*. Cambridge, Mass.: Harvard University Press, 1981. (A classic survey of Solar System astronomy updated to include results of spacecraft missions to the giant planets.)

New Windows on the Universe

Cornell, James, and Paul Gorenstein, eds. *Astronomy from Space*. Cambridge, Mass.: MIT Press, 1983. (Essays on the results of 25 years of astronomy from space platforms.)

Culhane, J. L., and P. W. Sanford. *X-Ray Astronomy*. New York: Scribners, 1981.

Eddy, John. *A New Sun: The Solar Results from Skylab*. NASA Report SP-402. U.S. Government Printing Office, Washington, D.C.: 1979.

Fazio, Giovanni. "Infrared Astronomy," in *Frontiers of Astrophysics*. Cambridge, Mass.: Harvard University Press, 1976. (Detailed discussion of infrared detectors and the goals of infrared research.)

Field, George B., and Eric J. Chaisson. *The Invisible Universe*. Boston and Basel: Birkhäuser, 1985. (An up-to-date survey of discoveries in wavelengths beyond the visible. Includes a discussion of the major astrophysical questions still remaining and how they may be addressed by new instrumentation. Illustrated with many unusual computer-processed images.)

Habing, Harm J., and Gerry Neugebauer. "The Infrared Sky," *Scientific American*, Vol. 251, November 1984. (A summary of results from the IRAS mission.)

Hanle, Paul A., and Von Del Chamberlain, eds. *Space Science Comes of*

Age. Washington, D.C.: Smithsonian Institution Press, 1981. (A review of the history and achievements of space science.)

Hirsch, R. F. *Glimpsing an Invisible Universe: The Emergence of X-ray Astronomy.* New York: Cambridge University Press, 1980.

Tucker, Wallace. *The Star Splitters.* NASA Publication SP-466. Washington, D.C.: U.S. Government Printing Office, 1984. (Results of the NASA series of High Energy Astronomy Observatories.)

Tucker, Wallace, and Riccardo Giacconi. *The X-ray Universe.* Cambridge, Mass.: Harvard University Press, 1985.

New Instrumentation and Future Research

Astronomy Survey Committee. *Astronomy and Astrophysics for the 1980s.* Washington, D.C.: National Academy Press, 1982. (Volume I of the "Report of the Astronomy Survey Committee" lists the national priorities for research in the next two decades, with descriptions and estimated costs of proposed instrumentation. Excellent section on "The Frontiers of Astrophysics.")

Bahcall, John, and Lyman Spitzer. "The Space Telescope," *Scientific American*, Vol. 247, July 1982.

Burbidge, Geoffrey, and A. Hewitt, eds. *Telescopes for the 1980s.* Palo Alto, Calif.: Annual Reviews, 1981.

Field, George B., ed. "Special Issue: New Instruments for Astronomy," *Physics Today*, November 1982. (Includes "High-Energy Astrophysics," by George W. Clark; "Facilities for U.S. Radioastronomy," by Patrick Thaddeus; "Ultraviolet, Optical, and Infrared Astronomy," by E. Joseph Wampler; "Contemporary Planetary Science," by Michael J. S. Belton and Eugene H. Levy; and "A Golden Age for Solar Physics," by Arthur B. C. Walker, Jr.)

Frost, Kenneth J., and Frank B. McDonald. "Space Research in the Era of the Space Station," *Science*, Vol. 226, December 21, 1984.

Ryle, Martin. "Radio Telescopes of Large Resolving Power," *Science*, Vol. 188, June 13, 1975.

Stachnik, Robert, and Antoine Labeyrie. "Astronomy From Satellite Clusters," *Sky and Telescope*, March 1984.

Traub, Wesley A., and Warren F. Davis. "The COSMIC Telescope Array: Astronomical Goals and Preliminary Image Reconstruction Results," *Proceedings of the Society of Photo-Optical Instrumentation Engineers*, Vol. 332, 1982.

Waldrop, M. Mitchell. "Space Astronomy: The Next 30 Years," *Science*, Vol. 218, December 10, 1982.

Weekes, Trevor C., ed. *The MMT and the Future of Ground-based Astronomy*. SAO Special Report 385. Cambridge, Mass.: Smithsonian Astrophysical Observatory, 1980. (The proceedings of a symposium marking the dedication of the Multiple Mirror Telescope in 1979, which set the stage for the development of the large optical telescopes planned for the 1980s and 1990s.)

Technology, Science, and Discovery

Cornell, James, and Alan P. Lightman, eds. *Revealing the Universe*. Cambridge, Mass.: MIT Press, 1982. (A collection of essays on the complementary roles of theory and experiment in astrophysical research.)

Harwit, Martin. *Cosmic Discovery*. New York: Basic Books, 1982. (An influential survey of the history and process of astronomical research that argues convincingly that the most important discoveries have often been unexpected, unrelated to mainstream scientific thought, and made by "outsiders." Elegant presentation of evidence for the theory that major discoveries follow major technological innovations in observational instruments.)

McKelvey, John P. "Science and Technology: The Driven and the Driver," *Technology Review*, January 1985.

Price, Derek de Solla. "The Science/Technology Relationship: The Craft of Experimental Science, and Policy for the Improvement of High-Technology Innovation," *Research Policy*, Vol. 13, 1984.

Index